4 WEEK LOAN

information store

221 972

AVIATION INFORMATION MANAGEMENT

Aviation Information Management

From Documents to Data

Edited by

THOMAS L. SEAMSTER
Cognitive & Human Factors, USA

BARBARA G. KANKI
NASA Ames Research Center, USA

Ashgate

Published by
Ashgate Publishing Limited
Gower House
Croft Road
Aldershot
Hampshire GU11 3HR
England

Ashgate Publishing Company
131 Main Street
Burlington, VT 05401-5600 USA

Ashgate website: http://www.ashgate.com

British Library Cataloguing in Publication Data
Aviation information management : from documents to data
    1. Airlines - Management 2. Information resources management
    3. Aeronautics - Data processing
    I. Seamster, Thomas L. II. Kanki, Barbara G., 1948-
    387.7'0684

Library of Congress Control Number: 2002100903

ISBN 0 7546 1966 4

Printed and bound in Great Britain by MPG Books Ltd, Bodmin, Cornwall

Contents

Introduction

Part 1 Structure of Aviation Operational Information

Part 2 Management of Aviation Operational Information

Part 3 User Innovations in Aviation Operational Information

Summary and Recommendations

List of Figures

List of Tables

List of Contributors

Robert K. Bouchard is the Totally Integrated Technical Aircraft Network (TITAN) Project Engineer with the Avionics Engineering group at Federal Express (FedEx). He earned an Avionics Electrical Engineering degree in Holland and a second Bachelors degree and MBA with Embry Riddle. He started work at FedEx as the MD-11 expert in the Maintenance Control group and then transferred to the Avionics Engineering core group as the project engineer for the Central Fault Display System (CFDS) and the On-board Maintenance Terminal (OMT) system, now TITAN. TITAN ties a number of aircraft network components into the FedEx ground network.

Gary Cosimini is Business Development Director of Cross Media Products for Adobe Systems. He joined Adobe in 1992 to work on the development and promotion of Acrobat and the Portable Document Format (PDF), helping pioneer the use of PDF in the prepress industry. Prior to joining Adobe, Mr. Cosimini was Senior Art Director at The New York Times, helping to create the Science Times section and introducing digital technology to that newspaper chain. In 1986 he shared a Pulitzer Prize with the staff of the Science desk for a series on the Strategic Defense Initiative, better known as 'Star Wars'.

Brian L. Coulter is Director of Flight Standards at JetBlue Airways. He has 22 years of aviation experience as a pilot. Over 15 of those years were with the Royal New Zealand Air Force that included two projects requiring regulatory and SOP development for flight operations and training. He has served as Director of Safety for Air South Airlines and Manager of System Control at WinAir, Inc. where he was responsible for manual development and certification. As Director of Flight Standards, JetBlue Airways, he has authored the Manual On Manuals supporting the Electronic Flight Bag.

Jack W. Eastman is the Senior Director of Standards, Quality Assurance and Development for Flight Operations at Atlas Air. He holds a Ph.D. in corporate strategy from Kent State University. He has been the leader in the creation and development of Atlas Air's innovative electronically-based, corporate-wide, integrated document services project. Prior to Atlas, Captain Eastman spent many years at Trans World Airlines as Managing Director of Flight Training and

Development, where he led the development of network and web-based deployment of flight manuals, training materials and computer-based training curricula.

Anita Kanakis is an IT Project Lead at United Airlines. Her twenty-plus year career in publishing began with a focus on typography and design. She has experience in implementing publishing systems and providing design specifications for corporate, legislative, educational and other state agencies. She has helped lead the early effort to automate publishing at United Airlines, providing infrastructure specifications for the reengineered publishing process, automated workflow and data conversion as a key member of the development team. She is currently expanding the scope of automated workflow and data conversion to identify cross-divisional solutions for sharing and optimizing information.

Barbara G. Kanki is a Research Psychologist at NASA Ames Research Center with a Ph.D. in Behavioral Sciences from the University of Chicago. She has conducted crew factors research in both aviation and space systems; from flight crew communications and Crew Resource Management, to crew issues in ATC, maintenance and space shuttle processing. Customer collaborators include NASA, the FAA, the National Transportation Safety Board, as well as airline, military, manufacturer and union organizations. She manages human factors research under several NASA safety programs, and has conducted FAA-sponsored research in airline operating documents. With Dr. Seamster, she co-chairs the NASA/FAA Operating Documents Group.

William W. LeRoy is Manager of Flight Manual Services with US Airways. He is a retired Air Force pilot with over 4,000 hours in fixed and rotary wing aircraft. At US Airways, he was the primary architect for the flight publications process, and created the first electronic library used by a major airline for pilot publications. For five years, he has chaired the Air Transport Association's Digital Display Working Group, which has been instrumental in defining industry needs and shaping the regulatory environment.

Thomas L. Seamster is Senior Research Scientist at Cognitive & Human Factors. He has directed research in aviation human factors of air traffic control, crew systems and crew training. He has conducted research on expertise in computer programming, flight crew resource management assessment, fighter pilot weapons deployment and spacecraft control. He has developed and tested user interfaces for both military and commercial systems and has developed expert and intelligent tutoring systems for the aerospace community. Recently, Dr. Seamster has conducted research in airline operating documents and, with Dr. Kanki, has co-chaired the NASA/FAA Operating Documents Group.

Terry J. Snyder is Manager of Publishing Standards at the United Airlines Flight Training Center. He has been using the publishing stakeholder perspective to help develop a data-driven publishing business process. He has also worked with Information Services and data stakeholders to ensure data is optimally structured, presented and managed for maximum accuracy and reusability. In his current position, he is responsible for ensuring the flight operational information utilized by United Airlines' crews, instructors and support personnel is consistently created and processed according to documented and approved Best Practices standards.

Ron A. Sorensen is President of I.T. Werx in South Surrey, B.C., Canada. Mr. Sorensen is an accredited Information Systems Professional (I.S.P.) with over 30 years of airline information systems management experience in planning, development, training, implementation and support. He has been instrumental in the merging of seven airlines and their associated systems, data, procedures and documents. Currently working in association with the Air Transport Association (ATA), Mr. Sorensen chairs the ATA Data Model Working Group and authors several chapters of ATA iSpec 2200. Mr. Sorensen is involved in establishing the Technical Information and Communication Committee's vision, mission, objectives, strategies and goals.

Norman E. St. Peter is Manager of Fleet Operations Support with American Airlines. His career with American spans over thirty years. He holds a degree in aeronautical engineering and is a rated pilot and flight engineer. He has worked in both the Flight Training and Flight Operations Technical departments at American. For the last eight years, he has managed the Fleet Operations Support group, which is responsible for all of American's flight crew operating manuals and associated aircraft operating document resources. In his capacity as manager, he has been responsible for the structure, format, style and content of these documents and resources, and is now developing plans for moving American to an electronic flight bag concept.

Ronald J. Thomas is the Supervisor of Flight Training with US Airways. He has been employed by US Airways for 14 years and is also a Captain on the Airbus A319/320/321 aircraft. Captain Thomas has assisted in the revision and maintenance of the US Airways flight document system for 12 years and has presented US Airways document system issues to the FAA, NASA, ATA and numerous industry groups. Captain Thomas was a contributing author to the NASA/FAA 'Developing Operating Documents, a Manual of Guidelines' publication.

Rick W. Travers has been Manager, Technical Writing and AOM Support at Air Canada. He is the Chair, Flight Operations Working Group (FOWG), ATA Technical Information and Communications Committee. As Manager, Technical Writing and AOM Support, Captain Travers has been responsible for the re-structuring and implementation of an integrated flight operations information schema at Air Canada. As the Chair of the ATA FOWG, he has championed common information standards and the development of their specification. In that capacity, he has promoted the use of an industry Phase of Flight Specification and the ATA System Specification for flight operations information data tagging.

Daniel R. Wade is Director of Business Development with Astronautics Corporation of America. Mr. Wade has assisted in the product development and market planning of many advanced computer and display products in his 18 years with Astronautics, including many US Government and internally funded research projects. He has led the internal developmental planning of ruggedized smart display systems for transport aircraft, airport emergency vehicles, ships and other vehicles. Mr. Wade is managing the ongoing effort for the Astronautics Pilot Information Display and is the coordinator of the Astronautics/Airline/Industry/US Government Electronic Flight Bag working group meetings.

Foreword

It is axiomatic that one should not confuse activity with accomplishment. Aviation, however, has stepped into the trap. A clear example of this: the proliferation of event reporting and data collection systems. Today, airlines collect loads of data. How much information is actually extracted from the data remains open to debate, nevertheless, airlines feel secure in the belief that archives mean prevention.

The goal of dealing with human error in aviation operations underlies the proliferation of data collection systems. It is noteworthy, however, that aviation collects data on mis-managed events leading to undesirable outcomes and failed human performance, exclusively. Therefore, it is valid to question whether the data has allowed aviation to better understand (a) the *processes* leading to undesirable outcomes, (b) human contribution to successes and failures in the aviation system, and (c) strategies for capturing successes and dealing with failures.

There is nothing wrong with data and archives, but I will argue that an inertia has developed by which aviation continues to collect data simply because we can afford to do it: computers are cheap. I will further argue that there is little to be gained by the inertial collection of the quantitative/descriptive data we have collected over the last twenty years. Such data will not tell us anything that we already do not already know. I will treble my argument by proposing that we need qualitative data that will inform us about strategies to deal with the residual risk inherent to the almost perfect level of safety achieved by contemporary aviation.

It is also axiomatic that aviation lives and dies by slogans which tend to cloud thinking, a point that I have discussed elsewhere. Therefore, we believe that 'safety is first', that 'human error causes 70 percent of accidents', and so forth. We also believe in the 'big picture'. Since our induction into aviation, we are taught to always keep the 'big picture' in mind. Allied to the notion of the 'big picture' is what we might call the 'big bang': breakdowns in aviation are the product of major failures and terrible blunders, and therefore require drastic solutions. Consequently, we set ourselves grandiose safety objectives, such as reducing accidents rates by whatever percentage in whatever span of time, we aspire to eliminate human error, we aim to eliminate CFIT/ALAR accidents, to eliminate loss of control and upset accidents, aborted take-off overruns, and so on and so forth. This is consistent with our 'big picture/big bang' approach: when facing a big problem, we invest big dollars into a big solution, and the problem goes away. Unfortunately, aviation safety is not that simple.

The 'big picture/big bang' approach is an ingrained component of aviation's professional culture. While appropriate in the old days, when aviation was a fragile system, it brings little contribution to the ultra-safe system we have managed to build. The days of getting big returns for our investment in safety dollar or of seeing immediate results are gone. Improving safety in the ultra-safe system demands heavy investments to obtain small improvements that will not be immediately evident.

We seem to be at a cross-roads, safety-wise. We might continue to do more of the unimaginative same with more intensity, in which case we will likely get more of the same frustrations. Alternatively, we might accept that contemporary safety demands fundamentally different, intelligent avenues of action. Such avenues should be based upon the conviction that we need informed strategies that aim at draining existing swamps in aviation where human error feeds. These strategies will not bring the kind of clear and present safety dividends that we are used to, but their continued development and perseverant application will, in the long run, define a context resilient to the slings and arrows that continue to pierce the armour of today's aviation.

The subject of this book, operational documentation and information management, is a perfect example of an informed and intelligent attempt at the latter alternative, one that will not provide a 'big bang', but that holds the potential to drain the swamp. Its editors and contributors join the camp of those who believe in the need for a fundamental shift in safety thinking, and in our perspective on aviation human factors. The shift in perspective, beautifully captured by this book, recognises that we can not simply continue monitoring undesirable outcomes and collecting data on failed human performance. It denies the value of looking at human performance as if it took place in a vacuum, if we are to understand why people behave as they do in a particular aviation work domain. It recognises that human behaviour is a reflection of the environment in which people work. I suggest that the beauty of the book - in addition to its valuable contents - is that it 'thinks outside the box': this book delivers applied cognition in aviation. I can not stress, strongly enough, how hopeful I am that applied cognition will take aviation to the next level, safety-wise.

In a relatively short time, applied cognition has provided a different and realistic picture about the human contribution to aviation safety. Before, the slogan justifying aviation Human Factors was human error: the infamous '70% clause'. Today, we know that the notion of human error is a social construct of irrelevant safety value. Today we know that what aviation Human Factors does is provide the means to contain the mis-management of adverse events resulting from unanticipated interactions between organizations, people and technology.

Continuing with options at hand, then, we might try to perpetuate straight jackets: we may continue to naively believe that people have one best way of doing things while pursuing the aviation system's goals; we may continue to reinforce stereotypes; and we may continue to admonish people not to make errors. This, as already proposed, will likely bring more of the same. On the other hand, we may try to understand the cognitive compromises faced by workers in the field: we may try to study normal work situations; we may try to understand what people *really* do while pursuing the aviation system's goals; and we may tell people what to do *after* they make errors. This is, I believe, where the future lies. This is, precisely, what this book does.

Should we attempt to further aviation's health through normative/prescriptive approaches, through behavioural stereotypes, through regulatory straight jackets or through slogans, my forecast is a dubious future if not a predictable failure. Should we instead attempt to further aviation's health through formative approaches, through applied cognition, through flexible regulation and through sound business management practices, then I believe the chance of success is on our side.

At the International Civil Aviation Organization, the choice has been made. It is clear to us that the aviation system cannot be entirely pre-specified. Human intervention will render attempts at pre-specification limited in value. We nevertheless believe that a flexible normative framework is necessary, but the challenge lies in the realtime implementation of such a framework. We are ready to accept that, during implementation, deviations from the framework will take place. Therefore, we will focus our efforts in making sure that means—of which this book is a perfect example—are available to keep proper control of the process of managing such deviations, rather than in the deviations themselves.

In pursuing a deviation management approach to safety, this book becomes a valuable tool. Beyond its practical value, this book is living example of a belief dear to my heart: it is always the best option to face issues and tackle challenges, no matter how difficult, rather than to refrain from action, decrying the difficulties involved in it.

Please enjoy the book.

Captain Daniel Maurino
Co-ordinator, Flight Safety and Human Factors
International Civil Aviation Organization
Montreal, January 2002

Acknowledgments

The topics and contents of this book are the collaborative product of industry participants of the NASA/FAA Operating Documents Group. In 1997, when this Group was formed, most operating documents concerns focused on current paper-based document design issues and the organization of operating document systems. Even at that time, the Group was beginning to plan for the development of electronic libraries and electronic flight bags. As the transition to electronic documents progressed, operator involvement expanded and suppliers and other industry working groups became involved with the Operating Documents Group.

We acknowledge the contribution made by all NASA/FAA Operating Documents Group participants over the past five years. We take this opportunity to thank them and to encourage their continued participation and support.

Special recognition goes to those that made outstanding contributions. First, all the chapter authors are commended for making this information available in a timely manner. Their persistence is especially appreciated in light of the extreme workloads that many incurred after the recent terrorist attacks. Two of the authors, Ron Sorensen, President of I.T. Werx, and Rick Travers, Chair, ATA Flight Operations Working Group (FOWG), did much more than prepare chapters. Ron and Rick have shared their ATA and industry expertise with the editors of this book, and have championed industry cooperation in developing aviation information standards.

We wish to acknowledge the input from more than 20 operators and suppliers who have freely shared their innovations and lessons learned with the Group. In addition, several ATA working groups have collaborated including the Digital Display Working Group (DDWG), the Data Model Working Group (DMWG) and the Flight Operations Working Group (FOWG).

Eleana Edens (FAA AFS-230) has supported the NASA/FAA Operating Documents Group through FAA ARR-100, Office of the Chief Scientific and Technical Advisor for Human Factors. In addition, NASA Ames Research Center, Human Factors Research & Technology Division, in particular, Hester Coan and Kirsten Patanker, have provided further support.

This book grew directly out of NASA/FAA Operating Documents Workshop IV, in Tampa, Florida, during May of 2001. The material from that Workshop would have remained a collection of slide presentations had it not been for the encouragement of John Hindley, publisher of the Ashgate aviation series. It has been a pleasure working with John who makes book editing and publication an enjoyable experience. Our thanks also go to the editors at Ashgate who have kept this process on track.

Finally, we wish to emphasize that the contents of this book constitutes the voluntary sharing of information among aviation industry contributors. The information is not proprietary nor is it regulatory in nature. Rather it is the collaborative pooling of expertise from industry members dedicated to discovering the most effective way of managing and organizing aviation information as documents transition to data.

Thomas L. Seamster
tseamster@qwest.net

Barbara G. Kanki
bkanki@mail.arc.nasa.gov

Glossary and Abbreviations

AC	Advisory Circular.
ACARS	ARINC Communications and Address Reporting System.
acceptance process	FAA review in cases where FAA approval is not required (NASA/FAA, 2000).
AFM	Airplane Flight Manual.
aircraft system	ATA Spec 2100 for the numbered and named conventions of aircraft systems (ATA-FOWG, 2001).
airline	(See **operator**.)
AOLS	Airbus On-Line Services.
approval process	FAA approval indicating the FAA has evaluated and approved the document or information (NASA/FAA, 2000).
AQP	Advanced Qualification Program.
ATA	Air Transport Association.
ATA Data Model	Aviation industry structured representation of information/data interchanged between a supplier and an operator, promoted by the ATA.
attribute	Source of information about an element (St. Laurent, 1999).
aviation industry	Commercial aviation, including operators, suppliers and regulators.
AWIN	Aviation Weather INformation.
BIA	Business Interest Architects group (ATA).
business process	A method or procedure that can be automated and that is used in a business community (UN/CEFACT and OASIS, 2001).
CFDS	Central Fault Display System.
checklist	Formal list used to compare, identify or verify a group of actions or items (NASA/FAA, 2000).
cockpit	The forward part of the fuselage containing all the instruments needed to fly the aircraft (NASA/FAA, 2000).
common terminology	Standard terminology used within a particular industry, such as the aviation industry.

conformance   Fulfillment of a product, process or service of all requirements specified; adherence of an implementation to the requirements of one or more specific standards or technical specifications (UN/CEFACT and OASIS, 2001).

content management   Creation, processing and delivery of content (Goldfarb & Prescod, 2001).

conversion   Transformation of information from one medium to another (Arms, 2000).

corporate standard   Data or information standard limited to an organization or corporate entity.

CRM   Crew Resource Management.

CSDD   Common Source Data Dictionary (ATA).

CSET   Certification, Surveillance, and Evaluation Team.

CWIN   Cockpit Weather Information Network.

data   (See **electronic data**.)

data architecture   Depiction of the distribution and access mechanisms associated with data for one or more applications (FAA, 1999).

data confidentiality   When information is not made available or disclosed to unauthorized individuals, entities or processes.

data element   Basic units of information in the EDI standards containing a set of values that represent a singular fact. They may be single character codes, literal descriptions or numeric values.

data interchange   Exchange of structured data between information systems within and across organizations.

data management   Process of applying a standard methodology and readily accepted principles and practices to the creation, collection, storage, retrieval and conversion to usable business information of organizational data (FAA, 1999).

data model   Representation of the things of significance to an enterprise and the relationships among those things (FAA, 1999).

data modeling   Representing the underlying structure of the enterprise's data so it can then be reflected in the structure of databases (FAA, 1999).

data object   Data being operated on (transformed, digested or signed) by an application (W3C-XML-Signature Syntax and Processing).

data type   Type of data to be used to represent the content of a data entity (UN/CEFACT and OASIS, 2001).

database	Collection of data items that have constraints, relationships and a schema. A collection of interrelated files stored together, where specific data items can be retrieved by various applications (FAA, 1999).
DDPG	Dispatch Deviation Procedures Guide.
DDWG	Data Display Working Group (ATA).
delivery	Local distribution of documents or data. Not to be confused with data interchange (see also **electronic data interchange**).
digital document	(See **electronic document**.)
display format	Organization of different types of data in a display, including data characteristics such as labels, and other user guidance such as prompts and error messages.
DMWG	Data Model Working Group (ATA).
document system	Entire collection of operating documents and manuals organized to be used to specify and direct flight operations (NASA/FAA, 2000).
document	Set of formatted and related information, either paper or electronic.
DRM	Digital Rights Management.
DTD	Document Type Definition – A formal definition used to specify different types of valid documents (Goldfarb & Prescod, 2001).
EFB	Electronic Flight Bag.
effectivity information	Cycle information as defined by the ATA (ATA-FOWG, 2001).
EGPWS	Enhanced Ground Proximity Warning System.
electronic data	Structured data in a form primarily for computer processing (Goldfarb & Prescod, 2001).
electronic data interchange (EDI)	Automated exchange of predefined and structured data for business among information systems of two or more organizations (UN/CEFACT and OASIS, 2001).
electronic document	Set of linked and related data in electronic form for the primary purpose of presentation (Goldfarb & Prescod, 2001).
electronic document system	Operating documents and manuals organized and managed as data to be used to specify and direct flight operations (NASA/FAA, 2000).

eXtensible Markup
Language (XML)   Language designed to enable the exchange of data
                       between different applications and data sources on
                       the World Wide Web and that has been
                       standardized by the W3C (UN/CEFACT and
                       OASIS, 2001).
FAA   Federal Aviation Administration.
FAR   Federal Aviation Regulation.
FCOM   Flight Crew Operating Manual.
flight deck   (See **cockpit.**)
FMC   Flight Management Computer.
FOM   Flight Operations Manual.
FOQA   Flight Operations Quality Assurance.
FOWG   Flight Operations Working Group – a working
                       group under the Air Transport Association's
                       Technical Information & Communications
                       Committee.
generic resources   Designation for data that can be shared amongst
                       different users like maintenance, engineering,
                       manufacturers, flight operations and training.
IATA   International Air Transport Association.
ICAO   International Civil Aviation Organization.
ICIS   Integrated Crew Information System.
industry standard   Standard used across an industry such as the
                       aviation industry.
information creation   Developing or writing documents, operational
                       information or manuals.
information
  management   Applying a standard methodology to the creation,
                       collection, storage, retrieval and conversion of
                       organizational information (FAA, 1999).
information
  repurposing   (See **repurposing.**)
information set   Part of the data set used in information
                       interchange. The emphasis is on the information
                       rather than on the document (ATA-FOWG,
                       2001).
intelligent
  document   Document that is capable of dynamically
                       reconfiguring itself based on user input. All
                       information not applicable to the user requested
                       input criteria is eliminated from the displayed
                       document.

knowledge base	Information database that, in addition to containing data, captures and contains business experience.
LOSA	Line Operations Safety Audits.
markup	Structure data stored in the same file as the content (St. Laurent, 1999).
markup language	Language used to denote structure data and information about a document's content (St. Laurent, 1999).
MEL	Minimum Equipment List.
metadata	A special type of data that describes other data (Goldfarb & Prescod, 2001).
MMEL	Master Minimum Equipment List.
NASA	National Aeronautics and Space Administration.
NASIP	National Aviation Safety Inspection Program.
non-normal	Procedures or documents used in nonroutine operations where actions must be taken to maintain system integrity or to protect aircraft, crew or passengers from hazard (NASA/FAA, 2000).
OMT	On-board Maintenance Terminal.
open source	Software released with its source code, permitting code modification as well as non-proprietary technologies (St. Laurent, 1999).
operating documents	Cards, checklists, guides, handbooks and manuals generally prepared by the operator and used in performing operational duties (NASA/FAA, 2000).
operational information	Information used in the conduct of flight, ground or maintenance operations.
operator	Air carrier or operator engaged in domestic or overseas air transportation, including major, regional and cargo operators.
phase of flight	Standard stages that occur in most operational flights; to include preflight, taxi, takeoff, climb, cruise, descent, approach, landing and after-landing (NASA/FAA, 2000).
philosophy	High level view of how an operator conducts their business and all operations (NASA/FAA, 2000).
PID	Pilot Information Display.
POF	Phase of Flight.

policy | Written requirement established by an operator's management to be complied with by specified personnel (NASA/FAA, 2000).

procedure | Written sequence of actions and/or decisions prescribed by an operator (NASA/FAA, 2000).

QRH | Quick Reference Handbook.

REDARS | Reference Engineering Data Automated Retrieval System.

repurposing | Reuse of information where an application determines how the content is to be displayed, what content is displayed and how it can be accessed through the use of a markup language (Idea Foundation, 2001).

RTCA | Radio Technical Commission for Aeronautics.

safety-critical | Information affecting the safety of flight with 'time-critical' as one of its primary properties.

SGML | Standard Generalized Markup Language.

shared information | Information that can be used within or across aviation organizations because it is based on a common terminology and data standards.

SID | Standard Instrument Departure.

SMIL | Synchronized Multimedia Integration Language.

style manual | Guide that establishes formatting and writing standards to ensure standard writing style, terminology, use of graphics and formatting across documents (NASA/FAA, 2000).

supplier | A general term referring to manufacturers and vendors of aviation systems and equipment.

SVG | Scalable Vector Graphics.

tag | A component of markup languages used to specify element beginnings and endings (St. Laurent, 1999).

TAWS | Terrain Awareness Warning System.

TCAS | Traffic Collision and Avoidance System.

TICC | Technical Information and Communication Committee of the ATA.

time-critical
information | Information, such as that in non-normal procedures, that requires immediate access in the cockpit and a component of safety-critical information (NASA/FAA, 2000).

TITAN | Totally Integrated Technical Aircraft Network.

user interface	All aspects of information system design that affect a user's participation in information handling transactions (NASA/FAA, 2000).
value chain methodology	A methodology used to identify those activities performed by the organization that add value (Porter and Millar, 1985).
W3C	World Wide Web Consortium.
WebDAV	Web Distributed Authoring and Versioning.
workflow	The sequence of activities performed in a business which produce a result of observable value to an individual actor (UN/CEFACT and OASIS, 2001).
XML	(See **eXtensible Markup Language**.)
XMP	eXtensible Metadata Platform.

References

Arms, W.Y. (2000), *Digital Libraries*, Cambridge, MA: The MIT Press.

ATA-FOWG (2001), *Flight Operations Technical Information Interchange Master Document*, Version 7, Washington, DC: Air Transport Association

FAA (1999), *Metadata Repository Requirements Document, Version 1.0*, Washington, DC: Federal Aviation Administration.

Goldfarb, C.E. and Prescod, P. (2001), *The XML Handbook*, Third Edition, Upper Saddle River, NJ: Prentice Hall PRT.

Idea Foundation (2001), *Content Repurposing with FrameMaker+SGML and XML*, White paper prepared by the Idea Foundation.

NASA/FAA (2000), *Developing Operating Document: A Manual of Guidelines*, Moffett Field, CA: NASA Ames Research Center.

Porter, M.E. and Millar, V.E. (1985), 'How Information Gives You Competitive Advantage', *Harvard Business Review*, 63:4, pp. 149-160.

St. Laurent, S. (1999), *XML: A Primer*, Second Edition, Foster-City, CA: M&T Books.

UN/CEFACT and OASIS (2001), *Proposed Revisions to ebXML Technical Architecture Specification v1.0.4*, UN/CEFACT and OASIS.

W3C (1998), *Document Object Model (DOM) Level 1 Specification*, Version 1.0, World Wide Web Consortium.

Introduction

Chapter 1

Context of Aviation Operational Information

Thomas L. Seamster and Norman E. St. Peter

Introduction

Operational information management is at a crossroads as it sheds the remaining vestiges of its paper-based processes and moves through the uncharted domain of electronic data processes. The final outcome is not yet in full focus, but real progress has been made in the transition to electronic documents providing the aviation industry with the first steps in the right direction. This book looks at a combination of industry initiatives and airline successes that point to the next steps that operators can take as they transition to a fully integrated information management system. Although the route has not been fully identified, it is evident that the key to successful long-term efficient information management is industry-wide cooperation.

At present there are limitations on the industry's ability to achieve real efficiencies. Some operators have focused on pieces of technology without a long-term vision and direction. As a result, there has been only limited success. Without industry cooperation and a long-term vision, operators will end up with pockets of technology difficult to integrate into an efficient and comprehensive solution. This book outlines the near-, mid- and long-term initiatives that will bring about next generation electronic document systems.

The authors, along with the work of the NASA/FAA Operating Documents Group and the Air Transport Association (ATA) e-Business Unit, aim to promote a renewed, strategic vision for the creation, management and use of aviation information in the electronic data environment. Document system development and management needs to be integrated with engineering standards and training, and no longer be viewed as just a publications department that produces manuals and guides. Beyond integrating documentation within flight operations, this book looks toward the future where documentation across maintenance, ground and flight will be integrated into a corporate information system conforming to industry data standards.

Scope of Aviation Operational Information

Historically, documentation in flight operations, in-flight services, ground services, logistics, maintenance and engineering has tended to be implemented independently along divisional lines, especially in larger operations. When operators have more than one publication effort, there can be substantial duplication of effort and inefficient interchange of information between external entities such as suppliers and regulators, on the one hand, and the different groups internal to the operation, on the other hand. Figure 1.1 represents the confusing data interchange tangle that can result from multiple, independent information islands. The lack of cohesion and cooperation across different entities has resulted in limited information exchange within organizations even when it involves the same or similar data. Because of the different needs of internal entities, intra- and inter-divisional document and information repurposing and/or reuse of information where applications determine what and how content is to be displayed has been restricted. Further, this compartmentalization of the documentation process has hampered workflow efficiency and has delayed the implementation of the electronic interchange of information.

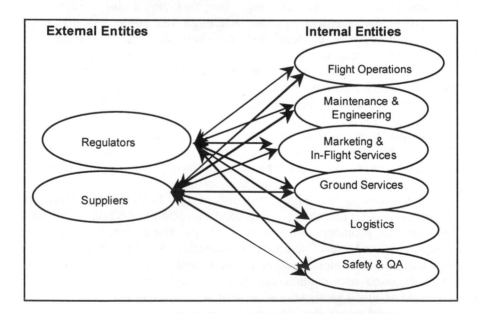

Figure 1.1 Current Information Islands and Interchange Tangle

The chief objective of this book is to outline the vision, planning and implementation needed to significantly improve efficiency of information creation, management and usage. The aviation industry currently has a number of disparate document formats and uses a range of different work processes within and between one or more of the following areas:

- Information creation by manufacturer and vendor, suppliers for flight operations, in-flight services, ground services, logistics, maintenance and engineering.
- Information management at the operator level for inter and intra divisional use.
- Information coordination and approval within operators and between operators and regulators.
- Information for training in-flight operations, in-flight services, ground services and maintenance.
- Information for use on the job in the cockpit, in the cabin, on the ground and for maintenance.

This management of information by different entities, with different formats to accommodate diverse user groups, has limited information sharing, linking and repurposing within and across aviation organizations. All these differences increase the complexity and cost of information exchange and reuse, making it difficult to automate document management functions.

Operational information does have some unique characteristics, but it also shares some common elements with maintenance. Both flight operations and maintenance can benefit from a better understanding of each others' information needs, with the goal of developing a set of common elements to facilitate information exchange, management and use. An important effort is underway at the Air Transport Association (ATA) where the Data Model Working Group (DMWG) is coordinating and modeling information requirements in the context of flight operations and integrating and harmonizing that model with their maintenance and engineering data in an ATA Master Data Model.

Toward a Standard Operational Information Interchange

The ATA Technical Information and Communication Committee (TICC), over the last few years, has been working on an industry standard for the interchange of information. This is not a regulatory standard where all will be required to use the same document and data format. This is an industry standard where operators can use formats of their own choosing to manage and exchange data within the organization as long as they are able to translate to and from that industry standard for interchange of data with suppliers. The DMWG, with other working groups under TICC, has been working on a data model to specify the primary entities with their complex interrelationships. Conceptually, the process is simple. Industry needs to account for both fight deck entities, such as phase of flight, as well as maintenance entities, such as aircraft systems, and combine these entities into a common framework for data interchange. Operationally, the challenge is enormous. An unprecedented level of cooperation will be required across the industry to agree on what needs to be standardized and how those standards will then be developed.

Collaboration Between Operators, Suppliers and Regulators

Work on industry standards has highlighted two factors that must be addressed in order to achieve an aviation industry information interchange standard.

The first factor is the essential interaction between operators, suppliers and regulators where each of the parties has been concentrating on their own requirements without considering the industry as a whole. Suppliers, operators and the regulator have their own set of needs:

- Suppliers are most concerned with the efficient and accurate production of documents in whatever form the different operators require without reference to standards.
- Operators, until recently, were most interested in converting manufacturer documents into their document formats as seamlessly as possible by using software compatible with that used by suppliers.
- Regulators have been most concerned with the approval process, often taking place at the document page level.

Each of these perspectives has different requirements with their own formats, processes and areas of emphasis. At present, the aviation industry, in collaboration with several ATA working groups, is developing standards for information interchange between manufacturers and operators. Suppliers have developed several different approaches to data interchange, and the regulators have their own information technology strategy (FAA, 1999). These efforts demonstrate some cooperation, but each has a different set of requirements. An industry information interchange standard will have to harmonize these different requirements. Organizations would use such an industry standard for converting their internal data when exchanging information between any combination of suppliers, operators and regulators.

Process to Product

The second factor is equally important and can be described as the process to product factor. This factor highlights the different requirements of those who manage documents, such as individuals in publications, and of those who work with the final product, the pilots and other end users. The transition to electronic documents provides an opportunity to ensure that improvements for one group do not result in degraded functionality for the other.

The document system managers have workflow requirements to automate and simplify document development, review, approval and dissemination. End users, such as pilots, have a different set of requirements based on using the documents in the operational environment. Both groups have usability issues that must be considered jointly so that the new electronic document system can be both manageable and operationally usable. This requires that both groups be active in developing an industry interchange standard with more than just token membership from one side or the other. It is important that industry not try to gain document management efficiency at the expense of operational usability.

Shaping the Standard

One example of a potential standard to be shared across operations is that of phase of flight (POF). POF is essential to the pilots in the cockpit with information requirements changing dynamically depending not only on the current phase but the previous and next phase as well. It is essential to account for cockpit informational needs within each phase of flight, but also in adjoining phases. Flight is not always linear. For example, climb does not always follow take-off. There are situations where a take-off is immediately followed by a rejected take-off which, although infrequent, has different informational requirements.

There are numerous POF listings developed by ICAO, IATA, ATA and others. From the perspective of each working group, their efforts are essential, somewhat unique, and at times, independent. From a broader perspective, these efforts are interrelated, and should all be heading to a point where there is one standard to which all others can be linked. The FOWG POF listing presented in Chapter 5 provides one of the most comprehensive listings that accounts for the varied sequencing of phases. This listing was published as an approved standard in ATA iSpec 2200 (ATA, 2001). This FOWG POF listing has the potential to become an industry standard across flight operations, in-flight services, ground services, logistics and maintenance and engineering.

Progress toward industry standardization may seem slow, but it is essential that industry work together to ensure an industry standard for information interchange. Once such a standard has been established, individual operators will have much better direction in selecting document formats and workflow processes that meet their internal needs while conforming with the industry standard.

Prioritizing the Issues

The issues addressed in the chapters of this book have been identified through four workshops over the past five years in a collaborative effort between industry and the research community. This effort, undertaken by the NASA/FAA Operating Documents Group, was started in 1997 to identify key operating document issues. The group has served as a vehicle for the operators to identify common needs relating to the best way to organize, present and manage information, especially critical information required for flight (Seamster, Kanki, and Coan, 2001). Although the Group has focused on the needs of flight operations across North America, its identification of issues and solutions proposed in these chapters are relevant to the international commercial aviation community.

The work of the NASA/FAA Operating Documents Group has paralleled that of the aviation industry as it transitions from paper documents to electronic data. The Group started with an initial emphasis on paper documents, with an interest in the transition to electronic documents. The Group, representing over 30 operators as well as aviation consultants and manufacturers, has identified two areas of growing importance to all operators, including majors, regionals and cargo (Seamster et al., 2001). These important areas are the organization of documents in the context of document systems and the transition to electronic documents or media. Issues of document organization have predominated, even with the Group's early work on primarily paper documents (NASA/FAA, 2000). But with the new concentration on electronic documents, issues of document system organization have undergone a shift in perspective and awareness. Operational information is no longer bound by the limitations of paper documents, and can be reorganized through a workflow process and accessed by the different user groups through browser technology. Within the Group, there has been a substantial shift from discussions about paper documents to the planning for and management of electronic data. With this shift from formatting paper documents to the management and use of electronic documents, the focus has also gone from looking at individual documents and files to looking at the content of those files as data.

The direction and focus has changed over the last five years, but there are still some fundamental issues that relate to electronic as well as paper documents. These are issues that cut across the different types of operators and which can be thought of as long lived concerns that require strategic solutions. These issues were identified through user feedback via surveys and ratings collected primarily during Workshops I and II (Kanki, Seamster, Lopez, Thomas, and LeRoy, 2000).

Overall rating results show that operators place an ongoing importance on issues about the organization of document systems. At the beginning, those issues focused on the reduction of the number of documents and merging and consolidating manuals (Kanki, et al., 2000). Overall, the most important issues concerned the organization of the document system and retrieving cockpit time-critical information. More recently, those issues of how to organize documents have combined with those of information architecture and management as operators are moving toward electronic document systems. Detailed operator ratings were collected in Workshop II (NASA/FAA, 1997). The top 15 issues are presented in Table 1.1 along with their mean ratings. These issues were rated on a five-point scale where 1 stood for Most Important and 5 for Little Importance.

Standards and Guidelines

Issues related to electronic document standards and guidelines had some of the higher mean ratings. Operators have been, and continue to be, concerned with the need for standards and guidelines in this emerging area. Standards are needed at the aviation industry level so that electronic document systems can be interchanged and managed with a guaranteed level of data integrity and reliability. Guidelines are needed that address the unique characteristics of electronic documents, including how to display the information and how users are to navigate. The transition to electronic data can bring benefits, but often at the risk of information overload and poor design (Lintern, Waite, and Talleur, 1999). New displays remove page size, weight and volume constraints of paper-based systems, providing developers with numerous user interface and information structure options.

Need for Analysis

Another top issue is the need for operators to perform their own analyses. Most current transitions, such as those in Parts II and III of this book, tend to be groundbreaking efforts without the benefit of existing guidance. Thus, extensive internal planning, prototyping and collecting of user feedback needs to be undertaken early on to ensure successful transition to electronic documents.

Table 1.1　NASA/FAA Workshop II Top Rated Electronic Media Issues (29 raters where 1 = Most Important and 5 = Little Importance)

Transition to Electronic Media Issue	Mean Rating
Take advantage of technology, don't use technology for technology's sake	2.04
User interface/usability issue (e.g., hypertext)	2.07
Industry guidelines needed (e.g. SGML) media	2.10
End-user benefits	2.14
Extensive analysis required	2.14
Security	2.17
Access/distribution	2.18
Desynchronization of paper vs. electronic	2.21
Limited applicable standards for electronic documents	2.28
Training cost reduction	2.29
Content guidelines remain same; presentation/format to optimize electronic media	2.32
Certification and approval process unclear	2.34
Revisions dates on electronic media	2.38
Printing and distribution savings	2.61
Weight/cost fuel reduction	2.68

Access, Distribution and Security

Another highly rated set of issues is related to the appropriate use of new technologies, access, distribution and security. The issue here is ensuring that the industry take advantage of the technology in appropriate ways. Early electronic efforts, whether in documentation or training, have looked at supplanting paper with 'page turners'. There is a growing awareness that this does not result in optimal use of electronic documents. Paper has different advantages than electronic documents, especially as one looks beyond individual pages to the entire document management system. A number of the chapters in Part 2 of this book address the need for a new vision and model in order to take full advantage of electronic documents.

These new technologies offer greater flexibility in accessing and distributing operational information, but with an increase in responsibility for authorization and data security. When developing an electronic document system, organization must address several security objectives related to the content, source and use of the information (Lindstom, 2001). Confidentiality guards the contents of a document so that unauthorized individuals cannot gain access or use. Content integrity must be ensured especially when documents undergo a number of processes and may be distributed in multiple forms. In addition, the content must be verified before document interchange or other processes are executed. Finally, the signature source must be verified as the valid entity responsible for the document. The aviation industry needs to achieve all these security objectives across the full range of electronic document interchanges and processes. Security of electronic documents is essential in a safety-critical document system. The confidentiality, integrity, authentication and signature verification must be combined with a set of operational procedures to ensure that the correct revision is being used. Also the document system must have safeguards so that the data cannot be tampered with nor inadvertently changed. Analysis of actual security incidents shows that problems in this area are due to deficiencies in the security policy or in its enforcement (NASA/FAA, 2000). Both the technological and human elements of safety and security must be carefully planned and implemented, two areas addressed in a number of chapters.

Cost/Benefit Considerations

Of a more immediate nature, operators need to account for cost/benefit considerations when planning and implementing electronic document systems. There are indications that standard electronic document and information systems can save money. The ATA TICC estimates that its efforts have reduced internal operator distribution costs, reduced revision delivery times and increased information accuracy. Compared with such general claims, those working with operating documents in the cockpit are concerned with specific savings, such as the reduction in fuel costs due to reduced weight in cockpit documentation. In order to justify and also to better direct this transition to electronic documents, operators and manufacturers should produce accurate and detailed cost benefits looking at the efficiency and potential savings across the entire operation for both the near and longer term.

Organizations should also look at longer-term savings that will be made possible by the new information technologies. They should consider the cost of document updates as well as the cost of developing and maintaining the software system used for managing and storing data. Operators will have to evaluate systems at the organizational level in order to fully realize savings in these areas. Another area where information management and use can produce savings is through the use of newer markup languages that allow the separation of content structure from specific display formats. By separating content structure from format, the organization can present the information in different operational contexts without having to modify or change the content, and through the use of master style sheets, can automate key formatting and display functions. Finally, by designing a document system as a comprehensive database rather than a collection of documents, organizations will see greater efficiencies in searches, in updating a piece of data instead of a complete operational document, and in the long-term repurposing of information.

Top Issues

By updating these top issues in the context of current and upcoming technologies, the aviation industry can plan and start to implement new document systems based on electronic data shared within and between organizations. The industry needs to ensure that the standards catch up with the technology, that more effective software environments and tools are developed, and that guidance and regulations address these new areas. This will require unprecedented cooperation between operators, manufacturers and regulators. The cost of this cooperation may be viewed as very high, but it is essential, and the long-term benefits to industry are much higher.

Achieving Shared Information

The concept of shared information goes beyond documents to include operational data both within and across organizations. Some of the more efficient operators have already accomplished a level of standardization within their organization, referred to here as corporate standardization. At one operator (see Chapter 9) flight operations and maintenance publications have been combined in order to expand the base of shared information and repurposing. At the industry level, maintenance has also achieved some level of inter-organizational standardization on such key information elements as aircraft systems. The use of such maintenance standards allows different organizations to communicate more accurately and rapidly based on a relatively simple industry standard.

Corporate and Industry Standardization

Corporate standardization has at least two levels within most operator organizations. At the lower level, each division within an organization, like flight operations, has different departments such as flight technical, flight standards and training. These different departments use some of the same information in their manuals, logs and training material. Organizations can start by identifying the shared information already in use as well as additional information that could be shared across departments. At the higher level within an organization, there are the different operations that share information about the aircraft, crews and logistics. This higher level includes flight operations, in-flight services, ground services, logistics, maintenance and engineering as the most obvious, but other areas including finance and marketing are also relevant. Again, these different operations need to review their data assets and identify areas of duplication and sources of shared information.

Industry standardization reaches across the aviation industry and must involve operators, suppliers and regulators. Such standardization is achieved at the industry level and must involve most industry players in order to achieve the efficiency outlined in Figure 1.2. An obvious benefit of such standardization can be seen by comparing Figure 1.1 with Figure 1.2. From an organization's perspective, data interchange would no longer be many to many, rather it would be from other operators, suppliers or regulators to the organization's centralized database. Such a vision of shared information requires both corporate and industry standardization based on industry-wide cooperation and intra-organization coordination.

Some think that the quick path to standardization is by converting all documents to a markup language, but this is misleading and only a partial solution. Markup languages like SGML and XML provide a useful vehicle or technology but they do not specify how documents are to be marked up. The corporate and industry standards are what will provide real functionality to marked up documents. Markup languages have undergone a tremendous evolution in the past ten years, and some of these languages may offer an excellent environment within which the aviation industry can manage its document overload. For example, XML, a subset of SGML, has been estimated to offer about 80 percent of SGML's functionality but with just 20 percent of SGML's complexity

(St. Laurent, 1999). Unlike some other markup languages, XML structured documents are in a machine readable form which allows for substantial automation of the data and document management process. The use of XML, which is starting to see wide acceptance, should also benefit from the development of applications and tools that will work together to facilitate data management.

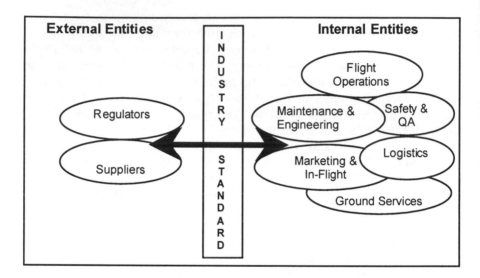

Figure 1.2   Electronic Information Interchange

Electronic Information Markup

Markup is also not the best way to model document elements and their relationships. Markup languages provide for document type definitions and schemas, but these are in text form and primarily for computer execution. Most in the aviation industry would find it difficult to understand the relationship between elements specified in a reasonably complex schema file. Those planning the development of electronic document systems to handle shared information should be working at the graphical modeling level which offers ways to specify data requirements in visually meaningful forms (Bird et al., 2000). Such efforts are currently taking place under the sponsorship of the Air Transport Association (see Chapter 5) and should be part of corporate planning as well. As the vision for shared information is being established through corporate and industry standardization, there is also a need for individuals who can serve as interpreters between the different user communities and the engineers working to develop applications and markup implementations. User input to this process is essential, but it must be informed input based on a good understanding of what the technologies can and cannot do.

Vision

A shared information vision for aviation is a complex undertaking, but technologies are starting to develop that can accelerate the process, and the potential benefits and long-term rewards can easily offset near-term effort and costs. One of the greatest problems is achieving cross-industry cooperation. Such cooperation is essential in developing the top-down standards, the security and the tools that will be required to manage such a system. Next to cooperation, there is a need for a better understanding of the current and future document management technologies, especially in the context of the quickly evolving world wide web.

Just as the authors of these chapters are developing new ways to model and manage documents and information, others in the industry need to identify shared operational information. It is important that management at all levels fully understand the possibility of shared information and the need for a coordinated transition from documents to electronic data. Not only operators, but the entire industry needs to start thinking about operational documents as data that needs to be managed and interchanged more efficiently.

Organization of the Book

Operational information issues facing the aviation industry fall into three categories based on the immediacy of their implications. There are a set of issues facing the industry that affect operations as they exist today. Then, there are issues with implications that extend out a year or two. Finally, there are issues with long-term implications going out five or more years. Those issues with near-term implications are in many ways the most tangible, and because of their relative immediacy, can seem the most prominent. These are the issues that may also produce the more immediate results. The tactical considerations are important, but they should be viewed in the context of the mid and long-term issues, those strategic considerations that will shape the future of aviation information. The long-term issues and their solutions are starting to provide that overall direction that the industry is still formulating, and can help develop a framework within which to consider the mid and near-term issues.

This book starts with the long-term considerations and works back to the near term. Part 1, Structure of Aviation Operational

Information, addresses the long-term concerns of the industry to establish a common strategic direction. The four chapters in this part provide both an operator and industry-wide perspective pointing the way to creating, managing and using operational information within and between operators and suppliers.

Chapter 2 by Kanki and Thomas establishes the basis for the book by describing the essential nature of operating document systems. They start by outlining the development process for paper-based and electronic document systems. This chapter highlights the operational safety-critical and time-critical requirements that must be used to guide the transition to electronic documents. These issues should be used as a basis for the extensive analysis required in a successful transition, issues that may get lost while grappling with the numerous technical challenges of moving to a new electronic system. The authors remind us that independent of technology, we should not lose sight of standardization, usability and other critical requirements of the operational environment.

In Chapter 3, Cosimini provides the current and longer-term view of document creation, management and use. He provides a brief history of the precursors to electronic documentation. He presents the state of desktop publishing, including current advances in document storage, delivery and display, and then examines the need for portable information throughout the operational environment. Cosimini helps to set a future vision where document publishing becomes integrated with an organization's network environment while addressing essential issues of security as access and distribution evolve with greater flexibility.

The core of Chapter 4 is an overview by Sorensen of the standards effort undertaken by the ATA and its Technical Communication and Information Committee (TICC). The overview starts with the recent history of aviation information standardization and its emphasis on maintenance, engineering and logistics. It describes these standardization efforts as they expand into flight operations and provides a future view of shared data. This chapter sets the stage for current and future work on industry standards addressing the need for industry-wide cooperation.

Chapter 5 by Travers focuses on flight operations standards as maintenance becomes more integrated with flight operations. The Flight Operations Working Group, under the umbrella of the ATA, has been concentrating on flight operations information exchange between suppliers and operators. This is an area that could return substantial efficiencies to the entire industry if agreement can be reached on common data standards. All operators must understand this effort as they move from the display of information to the manipulation of electronic data.

Part 2, Management of Aviation Operational Information, addresses the mid-term concerns in managing operational information. Chapter 6 by Snyder and Kanakis describes the process of moving from a large set of operational documents to an aviation documents database. The authors outline the steps required for optimizing information to achieve efficient management. They provide techniques for establishing working groups across the organization to manage shared data.

In Chapter 7, LeRoy presents some of the conceptual and technological considerations in moving structured information to the cockpit. The author provides an object model for organizing and displaying operational information to pilots in the cockpit. This model is used within a process to convert current documentation into electronic documents that can anticipate user needs in the context of phase of flight, flight tasks and other operational conditions.

Chapter 8 by Eastman provides the vision and practical considerations in establishing a shared information management system. A corporate vision must be established at the start as the core team is being formed. The vision needs to meet the requirements of the organization while remaining aligned with the broader efforts and standards being developed by the aviation industry. This chapter covers implementation issues along with important cost/benefit considerations.

Part 3, User Innovations in Aviation Operational Information, deals with the more immediate issues facing operators as they introduce electronic documents within operations. Chapter 9 by Coulter presents an innovative implementation of the Electronic Flight Bag (EFB) by a new operator. This chapter outlines the main planning and implementation stages from the perspective of a mid-sized operation. It also reveals some important lessons learned in this rapidly evolving area, including key unanticipated problems and their solutions.

Chapter 10 by Bouchard presents an integrated aircraft network, a key piece of the technology required for the realtime distribution and updating of an electronic documents system. The author explains the scope of the aircraft network, describing its key components. This chapter provides important insights into the operational improvements and cost/benefits achieved through such a network.

In Chapter 11, Wade presents the display of electronic documents in the cockpit. The chapter starts with an overview of display considerations when moving to an EFB. It then discusses the planning process along with implementation steps. It concludes with a summary of EFB cost benefits.

Chapter 12 by Seamster and Kanki summarizes the present and future of aviation information management. Throughout these chapters, the authors take an organizational and user-centric rather than a technology-centric approach, emphasizing shared information requirements and user needs that define the ultimate structure of electronic documents.

References

ATA (2001), *iSpec 2200: Information Standards for Aviation Maintenance* (CD-ROM), Washington, DC: Air Transport Association.

FAA (1999), *Information Technology Strategy, FY2000-FY2002*, Version 1.0, Washington, DC: Federal Aviation Administration.

Kanki, B.G., Seamster, T.L., Lopez, M., Thomas, R.J., and LeRoy, W.W. (2000), 'Design and use of operating documents', In Appendix A, *Developing Operating Documents Manual, Moffett Field*, CA: NASA Ames Research Center.

Lindstrom, P. (2001), 'Special Report: The Language Of XML Security', Network Magazine (June 2001), pp. 56-60.

Lintern, G., Waite, T. and Talleur, D.A. (1999), 'Functional interface design for the modern aircraft cockpit', *International Journal of Aviation Psychology*, 9, pp. 225-240.

NASA/FAA (1997), *Proceedings of the NASA/FAA Operating Documents Workshop II*: September 10-11, 1997, Dallas/Forth Worth Airport: American Airlines Flight Academy.

NASA/FAA (2000), *Developing Operating Documents Manual*, Moffett Field, CA: NASA Ames Research Center.

St. Laurent, S. (1999), *XML: A Primer*, Second Edition, Foster-City, CA: M&T Books.

Seamster, T.L. and Kanki, B.G. (2000), 'User-centered approach to the design and management of operating documents', *Proceedings of HCI-Aero 2000: International Conference on Human-Computer Interaction in Aeronautics*, pp. 151-156, Toulouse France, Cepadues Editions.

Seamster, T.L., Kanki, B.G., and Coan, H. (2001), 'Management of flight operating document systems', *Proceedings of the Eleventh International Symposium on Aviation Psychology*, Columbus, OH: The Ohio State University.

Part 1
Structure of Aviation Operational Information

Chapter 2

Operator Document Systems: Structural Tradeoffs

Barbara G. Kanki and Ronald J. Thomas

Introduction

In this chapter we review the main factors that currently determine the structure of operator document systems. Industry, corporate and operational requirements provide the fundamental criteria on which document system decisions are made. Although current experience in developing and maintaining documents is largely paper-based, many high-level requirements (particularly those related to operational needs for safety- and time-critical information) pertain equally to electronic and paper-based document systems. Further, these requirements tend to call out the defining features unique to information systems in aviation operations. Information access and usability in complex, dynamic and high-risk operational environments often drive the structural tradeoff decisions that are the topic of this chapter. While the exact form of the tradeoffs changes dramatically when paper documents transition to electronic media, many key determining factors remain the same. In the following sections, we will identify the requirements and limitations that impact the development of operating document systems and discuss the evolution of structural tradeoffs as paper systems become electronic. In focusing on structural changes brought about by technology advances in information systems, we will leave the more general guidance for developing and maintaining document systems to other document design references (Adamski and Stahl, 1997; Degani and Wiener, 1994; FAA, 1994; FAA, 1995; NASA/FAA, 2000).

Developing Operating Document Systems

The operator who is developing a new document system, or reorganizing an existing system, should review the entire document system as well as the complete operating documents process. That process includes not only the planning and organization for the document system, but the design, review, production, maintenance and distribution of system manuals and publications. Each part of the process will affect the entire system (NASA/FAA, 2000, Section 1.1.1).

Figure 2.1 illustrates the development of the operating documents process as a system. It also illustrates the types of processes that influence the way in which document systems are designed today. The decision process may not be implemented as explicitly or consistently as the figure depicts, but each element of the system, 1) organization of documents, 2) design of documents and 3) production and maintenance of documents is a critical part of every operator's document system. This development process applies equally well to the development of paper-based and electronic documents.

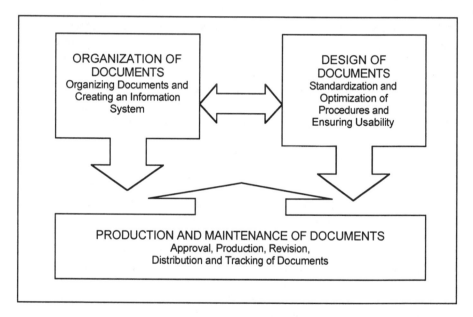

Figure 2.1   Operating Document Development Process as a System

The importance of clearly understanding and defining the three elements of document development as a system cannot be understated, and even at this conceptual level, one can anticipate how structural tradeoffs among system, document and production concerns can emerge. While safety-critical usability issues take precedence in some decisions, the elements must maintain a balance in order to function both smoothly and cost effectively. As an information management system that is dynamically changing, the development process is even more complex. Nevertheless, the obvious starting place is to identify the information requirements; those required through regulation, recommended by the industry, and established through corporate policy for operational purposes.

Regulatory Requirements and Guidance

The only requirements in the strict sense of the word are those required by law (e.g., the Federal Aviation Authority, The Civil Aviation Authorities). In the US, documents must contain information directly required by Federal Aviation Regulations (FARs). In addition to general requirements (e.g., FAR 121.131, 121.133, 121.135, 121.137, 121.139, 121.141, 121.315), numerous FARs contain specific information that must be included in an operation's document system. Some FAR parts to review are FAR 21, 23, 25, 61, 91, 121 and 135 as applicable. One way to ensure you have all information required by these regulations is to use the National Aviation Safety Inspection Program (NASIP) Checklist (FAA, 1996). This is the document the FAA uses to ensure compliance with all applicable FARs or Advisory Circulars (ACs).

The Air Transportation Operations Inspector's Handbook, 8400.10 (FAA, 1994), is used by the FAA to provide direction and guidance for certification, technical administration and surveillance of air carriers operating under FAR parts 121 and 135. As such, it also provides the operator with guidance on what the FAA is looking for. For instance, information on manuals is described in Volume 3, Chapter 15, Manuals, Procedures and Checklists. Operating specifications that are based on 8400.10 list the operations approved by the FAA. Some operators reproduce the complete operation specifications while others include only information that is pertinent to the flight crew.

Advisory Circulars (ACs) are additional sources of information applicable to operations. The title of the AC usually provides enough description to determine whether it is applicable and AC numbers start with the FAR part number they are associated with (e.g., 21, 91, 121). Finally, information contained in the Aeronautical Information Manual (AIM) (FAA, 2001a) and ATC Handbook (FAA, 2001b), while not required to be in manuals, may help clarify policies and procedures.

Supplier Information and Manuals

Suppliers, including manufacturers and other vendors, provide specific information on their products (aircraft, engines, individual systems, etc.) in flight manuals, operating manuals and flight training manuals. Operators should comply with supplier information and ensure that the resulting documents meet the needs of their user groups (see NASA/FAA, 2000, Section 2.1.2). For example, manufacturers provide basic information about the engineering of the aircraft and its subsystems, and the operator must ensure that information is compatible with their operational procedures and practices. FAR 121.141 allows a carrier to modify and create aircraft specific information to meet their operating requirements as long as it does not conflict with the manufacturer's information.

Other chapters in this book (see Chapters 1, 4 and 5) discuss the importance of supplier-operator standards and provide more detail on how such standards should be developed.

Operator Best Practices and Corporate Philosophy

Researching the document systems of other operators is of great practical value in redesigning a document system or developing a document system for a new fleet. Through a complete review of the document system and working agreements of an operator similar to your own, best practices may be identified. Required information may be placed in many different locations, so it is important to review the entire document system and not just primary documents.

In addition to learning from the best practices of other operators, it is important to embed the corporate philosophies and policies that form the basis for your own procedures and practices. With procedures forming a significant proportion of operating manuals, an integral process for developing procedures in concert with corporate philosophies and policies is key. Philosophies and policies should address the characteristics of the operational environment, so they can be used to develop relevant and usable procedures. Operators should identify or develop a consistent, high-level philosophy stating how the operation is to function. From this, other philosophy statements may be developed to specify essential aspects of the operation. Philosophy should not be a compilation of generic statements, rather, it should highlight the unique and most positive aspects of the operation's mission. Policies flow from philosophy and should be reviewed for consistency with other policies.

Organizing Criteria and Operational Priorities

Information Type Criteria

As information content is established, it must be organized into a document system that satisfies operational requirements when and where it is needed by various users. Because it is neither logical nor practical to place all required information into one document, grouping criteria which address the needs of your operation must be assessed in determining the contents of documents. Again, reviewing the document system of similar operators may help to identify primary groupings or information types. Useful information types include:

- Required in Flight.
- Aircraft Specific.
- Company Generic.
- Large Content.
- Route/Geographical.
- Phase of Flight.

These information types serve as practical grouping criteria. For example, information required in flight helps determine what must be included in the space- and time-constrained cockpit environment. The second information type, aircraft specific, may be used to determine how to separate information for each fleet in the operation. The Airplane Flight Manual (AFM) from the manufacturer will already contain a large portion of this information. Some operators use the manufacturer's AFM unaltered while others modify and reorganize the contents to suit their operations. Just as it is logical to locate aircraft specific information in a separate manual, operator generic information can also be placed in its own manual. However, combining aircraft specific and corporate information into one manual may be convenient for the user if the resulting manual is not too large. When working with paper-based systems, if an information item has large content, a separate document may be advisable, as in the case of the Minimum Equipment List (MEL). When working with electronic documents, document size constraints are replaced with user interface considerations, including locating information and navigating between documents.

Information pertaining to the route of flight or certain geographic areas is another information type typically grouped together for efficient organization (e.g., departure, arrival and approach charts as well as station information). When the operational routing structure is limited, however, only a subset of information may be required in flight. As a final example of information type, phase of flight is an essential organizing criteria both within individual paper-based manuals and for the entire electronic document system (see Chapter 5). For instance, information such as take-off and landing performance data must be grouped on the basis of the phase of flight in which it is used.

Phase of flight may also become one of the primary entities as operators convert information to electronic document systems (see Chapter 12).

Table 2.1 gives examples of several information types (see NASA/FAA, 2000, Section 1.4.1) that may result in grouping information in particular documents or locations because their operational purposes have shared attributes. Additionally, Table 2.1 demonstrates that many of the information topics may be used in more than one place. For example, training information may crossover into many other topics. Here, the concept of information redundancy becomes part of the document system decision process because the choice between referencing other documents versus including information wherever it is needed is a dominant issue in paper-based systems. Redundancy is handled quite differently in an electronic system or hybrid paper-electronic system, but the identification of common information for multiple purposes or user groups is important to determine for any document system. The repurposing of information is both a prime benefit and a challenge in developing a fully integrated information management system.

Information Importance and Information Users

When information is important to the operation, it should be easy and quick to locate. Major considerations include: 1) time constraints, 2) frequency of use, and 3) information users.

The most important are information requirements imposed by time constraints during flight operations (see NASA/FAA, 2000, Section 1.4.2). The main levels of priority include:

- Time-Critical – information is required in flight; can jeopardize the safety of the flight if not immediately available (e.g., emergency procedures).
- Time-Sensitive – information is required in flight; can affect the level of safety or delay the operation if not available in a short time period (e.g., crosswind/tailwind landing tables).
- Frequent – information is required in flight; does not fall under levels one or two.
- Home Reference – information is not required in flight.

Table 2.1   Types of Information Organized into a Document System

GENERAL	GENERAL POLICIES & PROCEDURES
Normal operating policies & procedures	Management structure & personnel
Non-normal op. policies & procedures	Authorized operations
Limitations	Operational control
Performance	Flight planning
Maneuvers	NOTAMS & PIREPS
Supplementals	Jumpseat
Techniques	Hazardous material
Systems information	Passengers
Weight & Balance	TRAINING
Adverse weather	Basic indoctrination
AIRCRAFT SPECIFIC (AS)	Ground training
Normal AS operating procedures	Flight training
Non-normal AS operating procedures	Emergency training
Limitations	CHECKLISTS
Performance	Normal checklist
Maneuvers	Abnormal checklist
Supplementals	Emergency checklist
Techniques	Supplemental checklist
Systems information	QRH Checklist
Differences	NAVIGATION CHARTS
Weight & Balance	Standard Jeppesen Charts
Adverse weather	Airport Specific information
Minimum equipment lists	WORKING AGREEMENTS
Fault reporting manual	Flight crew rest restrictions

These distinct levels of time constraint clearly dictate the level of accessibility and usability needed when making organizational decisions about the document system. In addition, the frequency with which information is referenced determines accessibility priorities; information that is referenced often must be easy to locate and access. Information of this type may be placed or repeated in guides, cards or checklists. For electronic systems, frequently used information should be accessible with the fewest keystrokes of user actions.

Finally, information users define an essential information grouping criteria. In addition to the prime information users (employees responsible for accomplishing an item), it may be necessary or helpful for other employee groups to access this information, particularly when their jobs coordinate with that of the prime users. For example, pilots need to know how and when to start engines, but the pushback crew must know when this is going to occur in their pushback sequence and what to do if an abnormal condition arises. As more specific information types (e.g., aircraft-specific normal procedures and flows pertaining to pushback procedures) are determined, the list of information users may be filled out in more detail.

Preliminary Document List, Location and Usability Considerations

Once information is grouped according to organizing criteria, it is useful to create a preliminary document system list. As documents are titled and content is specified, it becomes evident where information is shared by different users and where information is repeated for safety and reliability or user convenience. A mockup of each document is useful for assessing physical and content size.

The next consideration in ensuring that the preliminary decisions make sense is document location. The primary locations for operating documents are in the cockpit, in the flight bag and on the ground. On the ground can include the pilot's home or crew base and should consider other employees who coordinate their tasks with the prime users. Some criteria for deciding on document location are:

* Electronic vs. Paper.
* Required in Flight.
* Volume/Weight.
* Wear and Tear.
* Level of Information.
* How Often Revised.
* Responsibility to Ensure Document Availability.
* Responsibility to Ensure Pages are Correct/Up-to-Date.
* Aircraft Accessibility.
* Cost.

The electronic vs. paper criteria affects all the other criteria with different consequences for paper, electronic and hybrid systems. For instance, electronic documents reduce volume/weight in the flight bag, increase accuracy and increase timeliness, but may be cost prohibitive. Costs and availability of electronic documents are changing rapidly and should be reviewed on a frequent and regular basis.

In a paper-based system, only those documents necessary for flight should be on the aircraft since space is limited in the cockpit. Volume/weight, again more of a paper-based constraint, is a consideration that is sometimes specified by working agreements that limit the volume/weight allowed in the flight bag. But a general criteria for document location–whether electronic or paper–is wear and tear, which will almost always be highest in the cockpit.

It may be possible to split some information between the flight bag, cockpit and on the ground on the basis of what level of information is needed or not needed in the cockpit. For example, aircraft systems descriptions can usually stay on the ground if enough schematics and controls and indicators information is available in flight. An alternative is to put systems information in a bound volume in the cockpit since this information is fairly large and does not change frequently. An electronic system is able to avoid many of these tradeoffs; yet the distinction between what information can stay on the ground vs. what must be available in flight is still critical to know so that levels of accessibility in an electronic system are appropriately designed.

Another criteria for location of paper documents is frequency of revision. Documents that must be revised often increase the complexity and cost of having them in the cockpit. Thus, frequency of revision is a key criteria in cost justifying and developing electronic document systems. Furthermore, if the primary location for the document is the cockpit, a system will have to be developed to ensure it is onboard and that spare documents are readily available. When documents are located in the flight bag or on the ground, individual users of the document more likely assume this responsibility, thus reducing system complexity, but requiring more copies of the documents. Similar to document availability, ensuring documents are current and pages are correct may require a special system for cockpit documents, while ensuring currency and correctness of documents in the flight bag and on the ground usually falls to the user.

Aircraft accessibility is another important document location criteria. If all your aircraft pass through one or two hubs on a regular basis, locating documents in the cockpit may be the most efficient. If they do not, it may be more reasonable to keep documents in the flight bag. Cost is often the bottom line criteria and the costs associated with maintaining a document in various locations should be compared. For example, fewer copies are needed if placed in the cockpit but higher manpower costs are likely.

Cards, Guides and Checklists

Cards, guides and checklists can greatly increase usability for time-critical, time-sensitive and frequently used information. In addition, they have been found to aid the performance of complicated items (e.g., FMS Guide, Altitude Capability Card), serve as memory aids, and allow splitting off frequently used information from more detailed information that can remain at home or in other documents. The downside of cards, guides and checklists is that as they grow in number they can be more difficult to use in the cockpit and manage throughout the revision process.

Indexing and Redundant Information

> Indexing the individual documents and the document system is extremely important for the users. No matter how clear and well written your information is, it has little value if it cannot be found in a timely manner (NASA/FAA, 2000, Section 1.4.7).

In both paper-based and electronic systems, there are important generic considerations to keep in mind. For instance: 1) indexing within documents as well as a master index across documents, 2) indexing for abnormal and emergency procedures consisting of a single source index, if possible, that will indicate the exact location of the information (consideration should be given to 'tabbing' each page for direct access once the location is known), and 3) user input and testing in index design to ensure ease of access under time-constrained conditions.

Redundant documents or redundant information in different documents may be desired for safety and reliability purposes; for example, both pilots frequently carry the same departure, enroute and approach charts. In a paper-based system, redundancy conflicts with space limitations, so tradeoffs must be carefully assessed. If information is repeated everywhere needed, user convenience is maximized, but document size and cost are increased and the revision process is complicated. If information is listed once per document (or in one document only) and is everywhere referenced to that location, the size of manuals and potential for inconsistency is reduced possibly at the expense of clarity and user convenience.

These organizing criteria are not independent factors; rather, information importance, information type, location and users are often correlated with one another. Together with information requirements, they comprise the rationale for compiling information into flight operating documents such as those in Table 2.2.

Reviewing and Testing the Document System

Once a document system is developed or reorganized, it should be reviewed and tested by end users under realtime or simulated realtime constraints. In addition, other internal and external information user groups should be involved in the review and testing process. Philosophies, policies and procedures of all these groups should be properly integrated.

Table 2.2　Examples of Organizing Criteria for Flight Documents

Information Requirement	Information Type	Information Importance	Information User	Information Location
FAR 121.135 manual contents	Aircraft Specific	In Flight Time-critical	Flight Crew only	Flight Bag
FAR 121.407 training program	Route/ Geographic	In Flight Time-sensitive	Flight & Cabin Crews	On Aircraft
8400.10 Sect. 3	Training	On Ground Reference	Flight Crew	Home

The document system should be reviewed on a continuing basis; 8400.10 (FAA, 1994) recommends this be done every one to three years in a stable environment. A review should also take place after major events (e.g., mergers, acquisitions, rapid growth, downsizing) and after technology advancements. The review process is greatly simplified if a document database or structured listing is maintained. This allows retracing of initial decisions to determine whether they are still the best options, and becomes a systematic means of assessing and justifying the move from a collection of paper documents to an electronic document system.

Structural Tradeoffs Today and Tomorrow

The organizing criteria and priorities provide operational guidance for making structural tradeoff decisions. In spite of many shared requirements, limitations and information sources, operator document systems show great diversity across corporations. A variety of factors account for this including the following:

*   Documentation requirements are generic and allow for customization to meet different operator needs.

- Operators do not always review document systems of other operators in order to benefit from best practices.
- In the US, regulatory approval is at the regional level, not through a central approval system.
- Tailoring of documents from supplier to operator, operator to departments, departments to fleets, et cetera, generate changes that are not always tracked.
- Global changes are cumbersome within longstanding heritage systems.

Aviation system changes such as air traffic or airport procedures, new aircraft or aircraft upgrades, have demanded a system for rapid update of operating documents. In addition, improved, cost-efficient electronic documentation systems are becoming available while existing paper systems are becoming more expensive to maintain. In the effort to keep up with industry changes and new technologies, there is continued need to maintain communication with the user communities, to adequately review and test new procedures and to discover effective and economical solutions for delivering materials. In light of the diversity of existing document systems, industry needs a standardized approach for information interchange that facilitates managing information for export to users via multiple delivery options (see the arrows at the bottom of Figure 2.2).

Much of our current thinking and approach toward building aviation document systems has focused on one user knowledge base at a time. Thus, while industry working groups, researchers and developers of technical publications have made significant progress in identifying critical issues at the procedural and document levels, little guidance has been provided addressing the entire operating document system shown in Figure 2.1 across all information users and their shared information requirements. Rather, it has been typical to follow one line of development. For example, the development of a specific paper document for flight operations, thus considering a small subset of the information sources, management processes and users as shown in Figure 2.2.

As we move into developing information systems for the future, the new technologies associated with electronic media have not only affected the look and feel of operating documents, they have provided tremendous opportunity to take a more comprehensive systems perspective. In addition to streamlining the production and maintenance of 'documents', the potential is there to improve consistency within and across documents, within and across user communities and to move more easily from one delivery system to another when needs, requirements and improvements warrant it.

New Technologies and New Tradeoffs

In reviewing the organizing criteria that underlie current structural tradeoff decisions, many can be reevaluated in light of newly available technologies and approaches. For instance, 'Can an electronic document provide the information needed by users when and where it is needed?' Issues of information access change dramatically as the concepts of page and 'page-turning' lose their relevance. Although it is possible to design a display to look like a page and to make button pushes or touchscreens equivalent to turning pages, such constraints are unnecessary and inefficient.

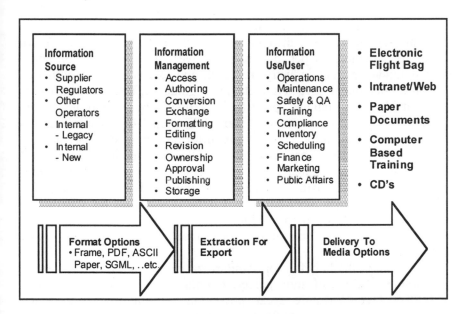

Figure 2.2   Potential Sources, Information Management and Users to Consider in a Publications Project

As electronic documents are constrained neither by size nor location in the same way as paper documents, considerations of stand-alone manuals and size limitations continue to change. With electronic document search capabilities and fewer size limits, indexing becomes crucial to success. A significant potential benefit of electronic media is ease of navigation through the use of electronic linking across what used to be multiple standalone manuals. However, pilot feedback and simulator testing with realistic scenarios will be key in determining whether these new systems meet user needs for timely access of operational information.

Standardization and Usability Issues

At a generic level, the issues of standardization and usability for onboard documents are the same regardless of media. For instance, it is always important to determine procedure flow, and maintain standardization across fleets and documents. However, at the procedure level, there may be many subtle differences. For example, consider a specific procedure that requires the pilot to accomplish one or more additional partial or complete procedures in order to complete the original procedure. Because the electronic document is not page-constrained, it could incorporate all relevant steps into a single procedure. This difference effectively reduces the number of unique procedures that must be trained and used.

At the document level, differences are numerous and obvious. With respect to standards, paper documents have the advantage of some external guidance from industry, research and experience. In contrast, electronic documents have relatively little guidance, and research often falls outside the unique priorities defined by aviation usage. Additionally, the concept of standardization of documents is particularly difficult when parts of the information system are paper-based and other parts are electronic. The document development process for electronic media has similarities to that of the paper document but not without major changes. For instance, the introduction of a new electronic document requires media acceptance and training in addition to procedure training, thus re-emphasizing the need for effective communication mechanisms to the user communities.

Nevertheless, the benefits for standardization and usability are great. For instance, the use of style guides, templates and software that can standardize layout acquire great power in maintaining consistency across fleets, across documents and across departments. Consistency can be checked and maintained with respect to format, terminology and indexing, as well as conventions pertaining to the actual accomplishment of the procedure (e.g., use of conditionals, decision trees). The document development process for electronic media requires greatly revised methods and standards (e.g., internal and external approval process, production process, communication mechanisms). But an effectively revised document development process will waste fewer resources and improve distribution and tracking, such as updating via remote uplinking, which will help to ensure consistency within the entire information system.

Summary

Many elements in the information system can be enhanced through new approaches related to electronic document systems; for instance, resolving issues of redundant information and cross-referencing, indexing and navigation within and across documents. In addition, issues of standardization can be resolved that go far beyond the document systems of specific user groups. New levels of standardization can be achieved that link user groups within operations, as well as operators with their information sources, such as regulators and suppliers.

Nevertheless, several issues remain regardless of media choice: the importance of incorporating feedback from all users involved, of testing under operational conditions, and introducing new materials through established communication mechanisms (Kanki, Seamster, Lopez, Thomas and LeRoy, 2000). In other words, while newly achieved benefits of electronic solutions are impressive, they do not substitute for current industry best practices in developing procedural content that satisfies operational requirements. However, through appropriate testing, implementation and training, the lifting of paper-based constraints can take us a great distance toward enhancing the accessibility of time-critical information in flight while increasing the overall accuracy and efficiency of the information management system. This type of 'safety' benefit from the end-user perspective generates a win-win solution for users, operators and industry partners.

References

Adamski, A.J. and Stahl, A.F. (1997), 'Principles of Design and Display for Aviation Technical Message', *Flight Safety Digest*, Volume 16: 1-29.

Degani, A. and Wiener, E.L. (1994) *On the design of flight-deck procedures*, (NASA Contractor Report 177642), Moffett Field, CA: NASA Ames Research Center.

FAA (1994), *8400.10 Air Transportation Operations Inspector's Handbook*, Volume 3, Chapter 15, Manual, Procedures, and Checklists, Washington, DC: Federal Aviation Administration.

FAA (1995), *Human performance considerations in the use and design of aircraft checklists*, Associate Administrator for Aviation Safety, Human Factors Analysts Division, Washington, DC: Federal Aviation Administration.

FAA (1996), *8300.10 Airworthiness Inspector's Handbook, Appendix 6*, Washington, DC: Federal Aviation Administration.

FAA (2001a), *Aeronautical Information Manual: Basic Flight Information and ATC Procedures, Change 3*, Washington, DC: Federal Aviation Administration.

FAA (2001b), *Air Traffic Control, 7110.65M, Change 3*, Washington, DC: Federal Aviation Administration.

Kanki, B.G., Seamster, T.L., Lopez, M., Thomas, R.J., and LeRoy, W.W. (2000), 'Design and use of operating documents', In Appendix A, *Developing Operating Documents Manual, Moffett Field*, CA: NASA Ames Research Center.

NASA/FAA (2000), *Developing Operating Documents Manual*, Moffett Field, CA: NASA Ames Research Center.

Chapter 3

Structure of Information in the Future

Gary Cosimini

The Historical Basis of the Digital Page

Publishing, the art of capturing and sharing knowledge, has a rich history that parallels the advance of culture and the development of logic and democracy. From the first evidence of pictographic writing and painting on the ceilings and walls of caves to today's palmtop devices with handwriting recognition and ubiquitous networking, the recording of information has served to increase the appetite for education and, at the same time, fueled the demand for yet better means of storing and conveying content. The invention of paper by the Chinese in the first century led to the printing press as surely as Gutenberg's moveable metal type in the 16th century inspired Mergenthaler's Linotype type-casting machine in the 19th (Mengel and Yutang, 1954). Hand presses, originally made of wood, were replaced by metal as casting technology became commonplace during the Industrial Revolution. While the medieval church preserved manuscripts by laborious hand-copying, the use of metals, which could be cast into complex shapes and could withstand great stresses, allowed type to be cast and reused and high-speed rotary printing presses to be manufactured (Seybold, 1984). Newspaper, magazine and book publishing industries, soon followed by commercial advertising, developed from these advances and prospered.

One of the greatest motivators of technical innovation throughout human history has been, for better or for worse, war. The fruits of struggle, or fear in anticipation of future conflict, have in our century provided us with the basis upon which waves of successful categories such as consumer electronics and personal computing have been built. Either through government funding of research or by declassification of wartime work, ideas and inventions that were spawned in the service of the military have led to the improvement of commonplace digital appliances and tools we use every day. The aviation industry was forged in that same crucible. The use of digital technology resulting from such work to capture and record words laid the foundation for electronic documentation. From messages carved as alphanumeric glyphs on clay slates to the keyboard invented for the typewriter, the cathode ray tube that portrayed blips on radar sets, the magnetic disk storage devices that record data just as wire recorders once preserved sound, and transistors, the logical offspring of amplifying radio tubes: all these byproducts of defense and aerospace research appear in the personal computer we take for granted today. Current conflicts continue to play an essential role in the development of more secure information systems.

Along with these commonplace technologies has arisen the assumption that information, once captured, should be retained. The advent of inexpensive data storage, recordable CDs and the Internet have fanned the desire to be able to find and access anything written or depicted since the beginning of time, a virtual library of endless knowledge, the perfect application for electronic documents.

Genesis of the Electronic Document

The transition of typography from raised lead 'hot metal' composition to 'cold type' was made possible by transistorized circuitry developed in the 1950s. Early electronic typesetting machines had film or glass photographic matrices bearing images of the letters and digitally triggered strobes that projected letterforms through complex lens systems onto rolls of photographic paper. Later variants used cathode ray tubes and stored digitized letterforms to perform more or less the same process more efficiently. The precise pinpoint beam of the laser, however, provided the real breakthrough. Combined with inexpensive memory chips that could store an entire page worth of binary dots, laser imagesetters introduced in the late 1970s were capable of exposing an entire page, images and all, in less than a minute. All that was missing was a uniform way to translate pages into bits.

My company, Adobe Systems Inc., was formed in 1982 by a pair of scientists, John Warnock and Charles Geschke, who foresaw the need to represent content symbolically and in all its manifestations using the personal computer. Their first intention, in fact, was to build a graphics workstation that could take advantage of the programming techniques they had learned over the course of ten years working on projects ranging from harbor pilot and space flight simulators to typography and outline fonts. The commercial breakthrough Adobe was to take advantage of was the desktop laser printer, a more economical descendant of the high-speed office copiers popularized by Xerox, where, ironically, Warnock and Geschke had met while employed at the

Xerox PARC research center in Palo Alto, California (Lammers, 1986).

While working at PARC, Warnock and Geschke were exposed to inventions that have become common fare today: Ethernet, the mouse pointing device, bit-mapped screen displays, point-and-click icons and the interactive graphical windowing display of modern operating systems. Frustrated by the slow pace at which Xerox could bring the fruits of their research to market, they departed to start their own company. With the inspired adoption of their PostScript printing language by Steve Jobs for the Apple LaserWriter in 1985, Adobe helped father what came to be known as the *Desktop Publishing* revolution.

Desktop Publishing: The First Wave

PostScript is a computer programming language optimized to describe printed pages. PostScript language programs may be created by the printer driver of an operating system, by a computer application or may be even written out by hand. These programs are then sent as files to a PostScript interpreter, or raster image processor (RIP), which converts the page description into a series of binary signals to the laser which then projects text and halftone dots onto either a charged toner drum or light-sensitive material. The beauty of this technique was that the authoring systems and the typographical fonts, which were stored as outlines, were decoupled from the hardware used to produce the print. This so-called 'device independence' allowed a variety of authoring solutions and printing systems to flourish, much as the standard architecture of the IBM PC created a series of competitors devoted to its manufacture.

Adobe published the language specification to help establish PostScript as a de facto standard, but licensed its first invention, the PostScript RIP, to numerous laser printer original equipment manufacturers; Desktop Publishing was based on the widespread availability of these devices. It allowed home and office users to set up their own page proofing 'factories', and almost overnight decreased the cost per page of manufacturing printed matter, and consequently increased the amount of printed information available to levels unimaginable a decade earlier.

From Print to Screen

The success of the PostScript page description language led to development by Adobe in 1993 of an analogous form optimized for viewing on computer screens. This was Acrobat and the Portable Document File format (PDF), and it answered a simple yet perplexing need: how to send completely formatted documents from one computer to another regardless of operating system, source application or fonts. Due to rapid advances in computer software and consequent obsolescence of the many application files in which data is stored, Warnock and Geschke reasoned that preserving final-form pages would become a critical need.

One of the most compelling components of the Acrobat system was called Distiller. At the time Acrobat was introduced, most of the proprietary typesetting systems in the world had been replaced by or supplemented with PostScript-compatible RIPs. As a result, practically every document printed in high-quality form also existed in PostScript. The Acrobat Distiller took these PostScript page descriptions and reformatted them in the highly-compressed PDF electronic paper format. This made it possible for publishers to automatically duplicate their printed content at marginal costs. At the same time, Adobe published a language reference manual for PDF, making it a candidate for acceptance by public standards bodies and government agencies. Its success was thus assured.

Given that any document in the world could be converted to PDF, a program to view PDF files could then be created to work on any operating system, and to date viewers have been written for DOS, UNIX, Microsoft Windows, Linux, Palm and Windows CE.

Formatted documentation is important when collaboration between individuals is necessary. A pilot and a flight mechanic, studying the schematic for an electrical console, should not have to be concerned with whether they are looking at the same page, or whether the symbolic font used on an instrument panel diagram translated correctly on his or her laptop, or when downloaded over the Internet. The ability to accurately publish any content, anytime, anywhere has become a requirement for documentation systems, and PDF has become a standard for doing so.

Today, PDF is the most widely accepted document format in use on the Internet, second in use only to HTML. Google, one of the main full-text indexing services on the Web, has already made 22 million PDF files searchable, according to The New York Times. The use of PDF has benefited from the rapid expansion of the Internet and Internet-based email, and its popularity begat interest in something new, the electronic document or e-book. If a document could be printed or read on-screen from the same file, it now made sense to add features to make it more easily navigable on a computer or to take advantage of the Internet and store hypertext links, interactive form fill-in fields, and sound and video. In successive years, these features have been added to the PDF format, which was updated so that third-party application and system developers could create and interact with this new form, the electronic document.

The Second Wave of the Digital Revolution: Internet Publishing

Years before personal computing became popular, the Department of Defense Advanced Research Projects Agency (DARPA) was funding a multi-university experiment in packet switching called ARPANET. This seminal sharing of information between dispersed computers led to what we know today as the Internet. The Internet is a vast, lightly regulated, transparent network of networks connecting millions of individual computer users around the world, funded and governed by cooperative means (LaQuey and Ryer, 1993). In part due to its genesis in academia, the Internet has served as the basis for a community that has grown at an almost staggering rate, metamorphosing to resemble nothing less than the purposefully chaotic synaptic wiring of a vast human brain. In fact, the Internet has become like the mechanical repository of universal knowledge and institutional memory boldly predicted in a 1945 essay by Vannevar Bush, the World War Two Director of the Office of Scientific Research and Development, which he called 'the Memex' (Goldberg, 1988). The Memex was an imaginary machine that would remember everything, images as well as text, and that would interact with its human partner by multiple means ranging from bone conduction to encephalographic traces; some weapons systems and aircraft navigation augmentation systems have already come close to fulfilling this prophecy.

However, the interesting part of Vannevar Bush's vision was reserved for libraries. Granted the means to record all forms of content, the ability to access and use such information instantly would be critical. The general need for universal access conflicts, however, with the special need for privacy. This struggle will become a theme for successive waves of the information revolution to come.

Standards are What Make Universal Access Possible

From what we have learned in almost a decade of pervasive connectedness engendered by the *Internet Publishing* revolution, we can deduce some characteristics that authoring and delivery system developers should embrace to make information sharing and repurposing practical (see Table 3.1).

Table 3.1 Electronic Document System Deployment Considerations

Information Characteristics	Considerations
It 'Wants' to be Nomadic	Rapid advances in display and storage technology dictate that content will be shifted from older devices to newer ones for the extent of its useful life.
It Should be Portable	There will never be a monopoly on display devices or technologies. It is shortsighted to try to bind information to a particular hardware device. Information needs good containers to ensure its complete and secure transport and to allow reusability.
Standards	Open systems philosophy and humanly readable markup should be preferred over proprietary technology to assure survivability over the long term. Standards should be supported whenever practical to improved repurposing and connectivity to alternate information repositories.
Appropriate Use of Standards	Authoring standards like eXtensible Markup Language (XML) are used to preserve author's intent and organization. Final form standards like PDF can preserve appearance, printability and editorial enhancements such as hypertext navigation aides, but XML tagging should be embedded whenever possible to facilitate repurposing.

Network Publishing: The Third Wave of the Digital Revolution

As the Internet spurred the interchange of content on a global level, another phenomenon has occurred, the need to simultaneously publish to multiple media and devices. When the publishing medium was only paper, it was relatively easy to accommodate commonly used paper sizes in page layout and word processing applications, and to provide paper trays of appropriate dimensions in printers. But with the advent of handheld devices and lightweight operating systems like Palm and Windows CE, and a future range of intelligent appliances, set top boxes with Internet access, touchscreen and infinitely configurable LCD displays for the glass cockpit of the future, the need for more highly portable, reflowable and interactive forms will be seen.

This third wave of publishing technology is known as *Network Publishing* (see Figure 3.1). Adobe, among others, is aligning itself with a new set of partners and tools to serve and support this category. It is our goal to help owners of content publish multiple formats, including sound and video, efficiently and to retain as much of the original intent as possible, even if that intent is structural and invisible to the naked eye.

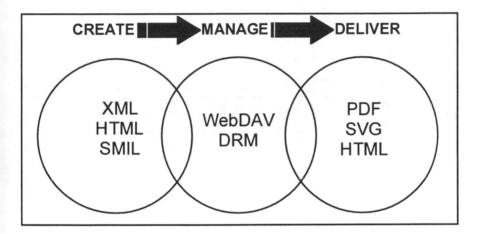

Figure 3.1   Representation of Network Publishing

Network Publishing requires improvements to current standards and will benefit from a new series of initiatives under development by members of the World Wide Web Consortium (W3C) and the Internet Engineering Task Force (IETF). Applications under development can then take advantage of these standards, resulting in improved authoring tools for PDF and other rich media. Network Publishing will utilize the mix of standards shown in Table 3.2 and described in the following subsections.

Tagged PDF

Since 1993, PDF has become a standard for electronic documents. It preserves the fonts, formatting, graphics and color of any source document, regardless of the application and platform used to create it. PDF files are compact and can be shared, viewed, navigated and printed as intended by anyone with free Acrobat Reader software. PDF is also supported by hundreds of third-party applications as well as the Macintosh OSX operating system.

Table 3.2   Formats and Standards in Support of Network Publishing

Standard	Functionality
Tagged PDF	XML structure and accessibility
XMP	Support for metadata, interoperability and extensibility
WebDAV	Support for metadata, interoperability and extensibility
DRM	Support for metadata, interoperability and extensibility
SVG	Support for dynamic, interactive graphics
SMIL	Support for dynamic, interactive graphics
OpenType	Cross-platform fonts
XML	Rules for tagging content

In 2001, Acrobat 5.0 introduced *Tagged PDF*, an enhancement that allows PDF files to contain logical document structure. Logical structure refers to the organization of a document, such as the title page, chapters, sections and subsections. Tagged PDF documents can be reflowed to fit small-screen devices and offer better support for repurposing content. They also are more accessible to the visually impaired. General information about Tagged PDF may be found by searching the Adobe Web site (http://www.adobe.com). Active PDF communities are located at the 'planetpdf' (http://www.planetpdf.com) and 'pdfzone' (www.pdfzone.com/webring/) Web sites.

XMP (eXtensible Metadata Platform)

Metadata–fielded information about files such as author, date and usage rights–is becoming increasingly important. Documents containing metadata greatly increase the utility of managed assets in collaborative production workflows. Adobe has just released version 1.0 of the eXtensible Metadata Platform (XMP) that provides applications and workflow partners a common XML framework that standardizes the creation, processing and interchange of document metadata across publishing workflows.

XMP embeds metadata inside application files. Because the metadata is enclosed within the file, documents retain their context when they exit their original system or environment. The embedded metadata can include any XML schema, provided it is described in Resource Definition Framework (RDF) syntax. Extensible, embedded metadata in application files provides significant potential for repurposing, archiving and automation in publishing workflows. Available as an open-source license, XMP can be integrated into any system or application. Adobe has integrated the XMP framework into Acrobat 5.0, InDesign 2.0, Illustrator 10 and future versions of Photoshop.

Industry support for the XMP framework has been announced by companies such as Documentum/North Plains Systems, IBM, Artesia, Xerox, WebWare, Kodak, KPMG and many other content repository and database asset management systems and workflow applications. Additional XMP information and a white paper may be obtained by searching the Adobe Web site (http://www.adobe.com). The W3C site (http://www.w3.org) has a resource definition format.

WebDAV (Web Distributed Authoring and Versioning)

WebDAV is a workflow and access control protocol based on HTTP extension that allows files on Web servers to be managed, versioned and is an open standard developed and supported by such industry leaders as Microsoft, Adobe, Cisco, Netscape, Novell, Apache and Oracle. It is a feature built into new releases of applications including Acrobat, InDesign, InCopy, Photoshop, Illustrator and GoLive. There are many uses for WebDAV, among them sharing files in a collaborative workflow and interchanging annotations between Acrobat 5.0 users. Additional information may be obtained by searching for WebDAV on the IETF Web site, and information about WebDAV and Acrobat may be located on the Adobe Web site.

DRM (Digital Rights Management)

DRM for Portable Document Format Files is provided by a trusted security and distribution system called Adobe Content Server. Content Server takes content created or archived in PDF and protects and distributes it directly from a Web site. The content is then securely locked to the purchaser's computer or disk medium in such a way that transfer is prohibited unless specifically authorized by the licensee. Content Server uses the 128-bit secure RSA encryption built into Acrobat viewers. Additional information on the Server may be obtained by searching the Adobe Web site.

SVG (Scalable Vector Graphics)

SVG is a World Wide Web Coordinating Committee standard nearing completion and represents a new graphics file format and Web development language based on XML. SVG enables Web developers and designers to create dynamically generated, high-quality graphics from realtime data with precise structural and visual control. With this powerful new technology, SVG developers can create a new generation of Web applications based on data-driven, interactive and personalized graphics. SVG has been incorporated in applications such as Illustrator, InDesign 2.0, GoLive and AlterCast, a server-based version of Photoshop. SVG examples may be found by searching the Adobe Web site and implementation information is available at the W3C site.

SMIL (Synchronized Multimedia Integration Language)

SMIL (pronounced 'smile') enables simple authoring of interactive audiovisual presentations. SMIL is typically used for 'rich media'/multimedia presentations which integrate streaming audio and video with images, text and other media types like SVG. SMIL is an easy-to-learn HTML-like language, and many SMIL presentations are written using a simple text-editor. The W3C site provides detailed information on SMIL.

OpenType

The OpenType font format is an extension of the TrueType font format, adding support for PostScript font data and developed jointly by Microsoft and Adobe. OpenType fonts are contained in a single file and are binary compatible across Mac and Windows. OpenType and its features are fully supported by InDesign and InCopy and by Microsoft Windows 2000 and later operating systems. The OpenType homepage may be found by searching the Adobe Web site. The OpenType font format addresses the following goals:

- Better support for international character sets.
- Smaller file sizes to make font distribution more efficient.
- Broader support for advanced typographic control.

XML (eXtensible Markup Language)

XML is a method for putting structured data in a text file. For 'structured data' think of such things as spreadsheets, address books, configuration parameters, financial transactions, technical drawings, etcetera. XML is a set of rules, guidelines and conventions for designing text formats for such data in a way that produces files that are easy to generate and read (by a computer), that are unambiguous and that provide extensibility, support for internationalization/localization and platform independence. XML is recognized, exported or used in the XMP metadata interchange format and in FrameMaker, InDesign 2.0 and InCopy 2.0, GoLive and Acrobat 5.0 for interchange of form field data. A good starting point for understanding XML may be found at the W3C Web site.

All of these open technologies and standards allow information to be shared, transformed and preserved. As the operating systems of hand-held devices and embedded displays become more advanced, Adobe plans to make its multimedia viewers available on as many of them as possible, with content optimized for each medium.

Security Considerations

Limiting use or access to electronic documents can be described by analogues to the physical world. There are two general approaches: physical, in that one may try to lock the 'door' to the place where the information is kept; and technical, through establishment of software-based trusted systems of encryption (Stefik, 1997).

One may draw a parallel from the history of religion. The prime cause of liberalization of the Christian church was the printed book, which gave the lay student access to what was previously held in handwritten manuscript form, chained to the wall of the monastic scriptorium. As books replaced manuscripts, the interpretation of sacred texts was opened to public debate and the authority of the official interpreters became subject to question; interpretation had been limited by securing access to the texts. The same holds true for electronic documentation, but locking the library doors is scant protection today and hardly feasible in the electronic context. Likewise, building proprietary digital formats and viewing systems is rarely practical from the point of view of cost. As standards for digital books, open systems and networking improve usability to information, the need to protect copyright and defend against mischievous use grows more urgent. These tendencies are contradictory and not easily reconciled, but the adoption of trusted repositories and licensing systems, with the benefit of ubiquitous network access to connect them to the user, may provide a system of control in the future. Some examples of trusted repositories, like Adobe Content Server, are emerging on the marketplace today.

All digital content is 'encoded'; that is, it is translated from one method or representation into another. For example, spoken words can be encoded into letters, and letters into a binary representation like ASCII which is stored in magnetized on-off 'flags' or binary bits on computers. When stored, transmitted or read into memory, these bits are available for interpretation by any computer programmer, and straightforward forms of encoding can be decoded easily by a hacker.

When someone wishes to protect these encoded symbols from eavesdropping or from surreptitious alteration, encoded information is re-encoded in a complex form called encryption. The degree to which encryption affords protection depends on the complexity and exclusiveness of the formula or algorithm chosen, as well as by the penetrability of the overall system of transactions itself. In general, given the reality of high-speed, cheaply available desktop computers that can test and retest combinations until some coherence appears, few systems are truly secure. However, some practical applications of secure forms using a combination of public and private 'keys' or passwords can provide sufficient protection for almost all uses. By adding features that provide a few more degrees of system communication and behavioral level, a great degree of protection may be achieved. Future forms of 'logins' will probably involve biometric or gestural recognition made possible by the same technology used to crack the simpler password of the past.

Implications for Aviation

The challenge of establishing systems of operational information management is one which must be shared and which must balance many contrasting needs: preservation of final form versus re-editable information, graphics versus text, security versus accessibility, and public standards versus proprietary computer applications. Like all engineering solutions, the ultimate answer will be a compromise that in twenty years may need to be changed again.

One of the early benefits of automation was thought to be the ability of machines to reproduce themselves, and indeed it was. The key factor in establishing a system of electronic document system should be its ability to 'reproduce itself' in different forms as the need and opportunity requires. This author's viewpoint is that the vision of Network Publishing coincides perfectly with that goal. The usefulness of information rests in its ability to be used whenever, wherever or in whatever form it will be needed in the future.

Applications of portable, mutable information in the aviation industry abound. Information systems that share content, from flight scheduling to billing and ticketing, with editable forms that can take input from a passenger's seatback LCD or a handheld device to make a reservation and select a seat using SVG graphic tied to an XML database, are just around the corner. Vector-based mapping systems displayed in the glass cockpit, with touch screens and dynamic update of routing or features, maps that can be printed, transmitted or stored for later recall, and whose coordinates relate precisely to onboard navigation controls, could be developed by instrumentation manufacturers using standard formats. Aviator's maps for the entire world could be stored in a handheld device and updated instantly from anywhere with Internet access. Service records could be evaluated automatically and digitally signed in a legally accepted form.

The risks of settling on any standard or solution, of course, need to be weighed against practical considerations such as commercial viability, availability of supporting services and applications, and universal acceptance. Technologies such as PDF, SVG and XML score well on all of these counts and hold great potential in the aviation industry.

References

Goldberg, A. (ed.) (1988), *A History of Personal Workstations*, New York, NY: ACM Press.

Lammers, S. (1986), *Programmers at Work*, Redmond, WA: Microsoft Press.

LaQuey, T. and Ryer, J.C. (1993), *The Internet Companion*, Reading, MA: Addison Wesley Co.

Mengel, W. and Yutang, L. (1954), *Ottmar Mergenthaler and the Printing Revolution*, Brooklyn, NY: Mergenthaler Linotype Company.

Seybold, J.W. (1984), *The World of Digital Typesetting*, Media, PA: Seybold Publications, Inc.

Stefik, M. (1997), *Internet Dreams: Archetypes, Myths, and Metaphors*, Cambridge, MA: MIT Press.

Chapter 4

Standard Aviation Information

Ron A. Sorensen

Introduction

This chapter provides an overview of why the aviation industry needs standards for operational information. It documents the evolution of the standards over three generations and includes a discussion of leading indicators such as vision, delivery, integration, retrieval and presentation. The background information explains the basis for the strategic vision that our industry must adopt, and the future capabilities provide the direction the aviation industry should be undertaking.

Why Standards?

If your business relationship with your customers/clients is as depicted in Figure 4.1 then you probably have no need for standards. You can agree with your trading partner what and how the information will be passed between you.

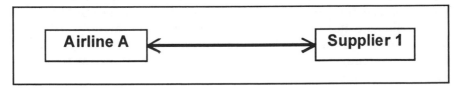

Figure 4.1 Is This Your Business Model?

However, I suspect your business relationship is more like that depicted in Figure 4.2. I would suggest you probably have multiple organizations you deal with and they also have multiple organizations they deal with. In which case I am sure you are discovering, or soon will discover, you cannot afford the time nor the costs associated with the negotiation, implementation, maintenance and support of multiple incompatible information interchanges.

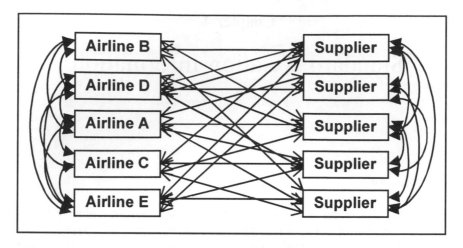

Figure 4.2   Doing Business 100s of Ways

One solution is to get together with other organizations that have similar interests. By pooling funds, new programs are started in areas where there are business needs, programs that otherwise could not be launched. This means that each participant's contributions become leveraged so that one dollar becomes two, three and even more when pooled with the funds of other partners.

This partnership approach to standards development is one that the Air Transport Association (ATA) has used for some time. ATA has already worked with both government and private sector organizations identifying business needs important to the aviation industry. ATA participants feel this is a cost effective and efficient approach to developing and delivering information standards enabling implementers to do business better while maintaining or improving safety.

Standards, in addition to enabling efficient data interchange, act as memory shortcuts that simplify user access to information and decision making. If a standard is known and trusted, users make decisions faster and more easily. Hence this makes a standard an extremely valuable asset. Standards also develop a sense of confidence and ease that may not be calculable in exact dollar terms. However, they have a definite monetary value. To complicate cost-benefit calculation even further, each operator and each supplier are at varying levels of automation and varying levels of process efficiency. Often process differences or inefficiencies can be traced to one or more company-specific sources (e.g., management/organization driven, union contract driven, physical location driven, financial approval driven, unique skill driven or legacy driven). General standards facilitate doing things better, cheaper, faster and in a more agile and flexible way. Standards facilitate improved information delivery, especially important when it concerns time-critical and safety-critical information. Additionally, standards can provide consistent, common and proven methods for data security.

Strategy and Direction

Over the past two decades, industry has experienced what could be considered as three separate generations of standards evolution. The First Generation comprised the period up to and including the 1980s. This generation focused on standards for the delivery of paper documents. Second Generation comprised the period up to and including the 1990s. This generation focused on standards for the delivery of digital representations of the paper documents. Third Generation, the current generation and still in an evolutionary stage, is focused on standards for the delivery of digital information as integrated/shared information using repositories, common terminology, a common data interchange and data security. The future direction and strategic vision for the Third Generation of aviation industry standards is a work in progress under the guidance and leadership of the ATA's Technical Information and Communication Committee (TICC).

First Generation - Up to and Through the 1980s

Leading up to the 1980s, the Maintenance & Engineering committees of the ATA worked primarily from a vision of paper documents (see Figure 4.3, column 1). The focus was primarily on establishment of standards for document size, paper size, weight of paper, font and everything to do with the presentation on paper. Focus also centered on establishing standards specifying what content would be contained in which manuals. During this period a group of ATA committees also focused on standards associated with the delivery of information. The delivery was paper pages and magnetic tapes. From an information integration standpoint, the manuals were stand-alone. There was no integration. The best one could hope for was a citation in a manual pointing to another manual. In most instances one relied on the user's knowledge and experience to guide them where to go and how to look for the information. From an information retrieval perspective, the best guide was the table of contents. There was no way to navigate directly to the lower levels of information. The presentation was static, and no matter what you did, the manual retained its original structure and format. As a result, the information retrieval was labor intensive.

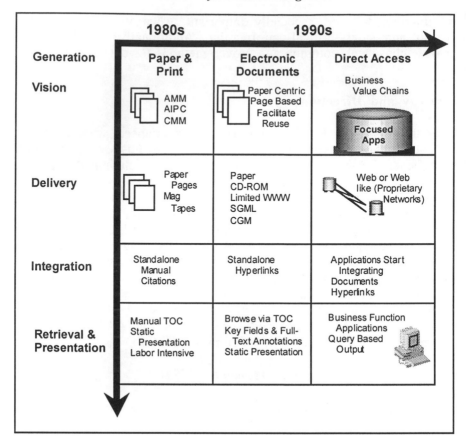

Figure 4.3   ATA TICC Past Strategy

Reality Byte: Putting it in Perspective

A paper Aircraft Maintenance Manual, when two side printed, is over six feet thick. In recent work with an American operator with five fleets, comprising some 230 aircraft, the amount of Jeppesen manual pages in active use by pilots totaled over five million pages. The operator allocated $34,000 US per year in fuel costs to carry the Jeppesen manuals on the aircraft. At the same operator, approximately 104,000 aircraft defects were reported per year. Each defect requires the mechanic to access information to rectify the defect. Each mechanic has access to approximately a million pages of maintenance data from which he must retrieve the maintenance procedures. This same operator also employed the use of over 18,000 rolls of microfilm cartridges, equating to over 44 million pages of maintenance information to support the operation. This adds up to an incredible amount of data, costs and time.

Second Generation: Electronic Documents and Direct Access

As we look at the next generation, the 1990s, electronic documents (see Figure 4.3, column 2), you notice we were still paper-centric and remained focused on pages. However, initiatives were being delivered to enable the reuse or repurposing of information. We utilized paper, CD-ROM, and initiated the use of the world wide web as a means of distribution. We were migrating to SGML for text and CGM Graphics delivery. Integration was still stand-alone documents; however, we were starting to use hyperlinks. Information suppliers were delivering text on SGML tapes with imbedded links (keys) to the actual CGM graphic contained on separate tapes. This enabled operators to retrieve and insert the correct graphic into the right information from the SGML tapes. Retrieval was done through a browser by selecting entries from the document's table of contents. There was the ability to index on key fields, to query on a documents text and to input and save annotations into the document. The presentation remained static. The document, was the document, was the document. The structure looked and felt the same every time.

During this generation some information suppliers initiated direct access (see Figure 4.3. column 3), providing operators the ability to directly access information from a supplier's repository. Utilizing the value chain methodology (Porter and Millar, 1985), the ATA Data Model Working Group (DMWG) analyzed the aviation industry, dissecting it into the business processes and product deliverables directly required to produce the end product. An industry's value chain model helps identify those activities that add value performed by the organization. The model may also indicate activities that do not add any real value.

ATA published this business value chain breakdown as part of the ATA Master Data Model. Data repositories were established focusing on specific areas of the business value chain. The delivery was web or web-like. However, the application and data were only accessible via a proprietary network. Although not based on industry standards, the information began to be integrated, or at least hyperlinks were used. Retrieval was provided via a proprietary business application that was typically a query-based output wherein you input your query and you get an answer. Boeing's Reference Engineering Data Automated Retrieval System (REDARS) and the Airbus On-Line Services (AOLS) are good examples of Second Generation proprietary direct access.

Strategy and Direction

We have moved from paper to electronic, and from there to direct access. Third Generation is moving to intelligent documents (see Figure 4.4, column 1). The major change here is the documents are in a database. Depending on your selection, the information is dynamically restructured, eliminating any information not applicable to your selection. This may represent a view focused on a specific subject, a view focused on a particular business function or many other alternatives.

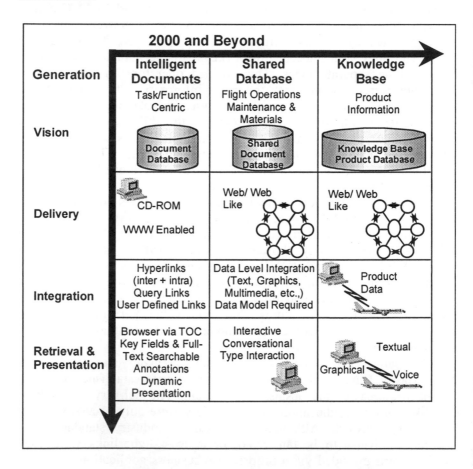

Figure 4.4 ATA TICC Current Strategy and Direction

Delivery is on CD-ROM and/or the world wide web. There is extensive use of hyperlinks within and now between documents. One can search and query across multiple documents. One can design and save their own custom searches. One can browse as in previous generations, but can do so now with full text searches. One can not only input and save annotations, but can also search annotations. There is also the capability to carry annotations forward from one revision to the next so that users do not lose their annotations as each revised document is received. And the presentation is dynamic. The ATA iSpec 2200 (ATA, 2001a) available on the ATA CD-ROM is an excellent example of an intelligent document.

Intelligent Document

Moving into intelligent documents adds shared database functionality (see Figure 4.4, column 2). This requires a major shift in the way we think, and the way we interchange, access and present information. Up to now, the focus has been on documents. Up until this point we have been giving users a library card and providing them with an exhaustive library. Users have had to know what document to go to. Nothing guided users to the document they required. Users knew that if they wanted certain information, they needed to go to a specific document.

A shared database moves the industry from documents to data. For this important transition, ATA participants have elected to proceed using an iterative development scheme based on a phased approach. The current focus is on supplier to operator interchange of flight operations, maintenance and logistics information. Once that area of interchange has reached a status of 'recommended for use' or 'approved for use', the business decides which of the remaining interchange categories is most important and beneficial to specify. This could be operator to supplier, operator to operator, operator to regulatory agency, supplier to supplier, supplier to regulatory agency, regulatory agency to operator, regulatory agency to supplier or regulatory agency to regulatory agency. This phased approach facilitates specification and implementation as business needs prevail. As each interchange category is initiated, the involved business units and information suppliers are enlisted and committed to producing the deliverables while fully aware of the potential savings and improvements they will enable.

The ATA TICC Steering Group has approved and published an interchange concept that outlines a vision and scenario of how this interchange will take place (ATA, 2001b). Initially this concept takes into consideration a supplier to operator interchange and accommodates the supplier's requirement to not allow operators direct access to their primary repository. As such, suppliers take an extract of what they want to release and place it on an external, secure access area. Operators either access it directly, if that is what they choose, or download and maintain a duplicate repository on their own servers (see Figure 4.5).

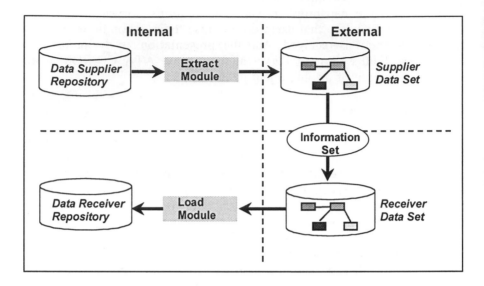

Figure 4.5 ATA TICC Data Interchange Concept

Information sets are constructed based on the ATA Data Model (ATA, 2001a). An information set is comprised of a consistent set of information that spans multiple business areas as represented in Figure 4.6. These sets are used to bind together shared information across documents as well as across organizational groups. Through the implementation of information sets, when operators incorporate a modification to an aircraft, and that modification results in information changes through various business areas, all appropriate information is updated and interchanged as an information set. The parts information, configuration information, maintenance procedural information, flight operations information et cetera, are all updated in a single information set. Since each change may impact different information, information sets are dynamic and contain all the changes required to ensure users maintain content integrity. This way, operators receive information cheaper, faster, with greater agility and flexibility while maintaining (if not improving) safety. Operators have the ability to implement information revisions utilizing processes that can be as automated and transparent as the operator desires.

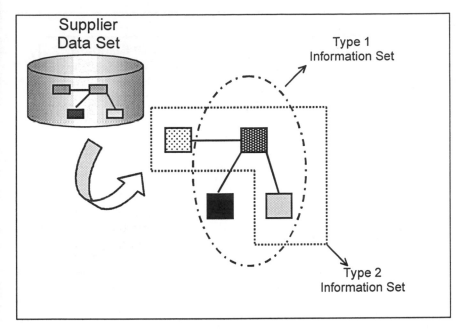

Figure 4.6 ATA TICC Information Set

There is data integration across text, graphics multimedia, color and 3-D. The ATA Data Model (ATA, 2001a) is utilized as a data model map. Using the ATA data model reduces instances of misinterpretation of the data. The data model specifies what data is kept about what objects that are important to the business. The data model also specifies the properties (lengths, etc.) and domain (permitted values) of the specified attributes. Additionally the data model defines the business rules in its relationships and their cardinality and optionality. Using the Data Model as a map facilitates the interchange of data without requiring suppliers, operators and regulatory agencies to utilize identical databases. Each organization can utilize their own proprietary applications and databases. However, they must translate their data into the agreed ATA data interchange standard. Once received, the information receiver, using the data model map (see Figure 4.7), translates the data into the receiver's database using their own mapping of the ATA Data Model to their internal data architecture.

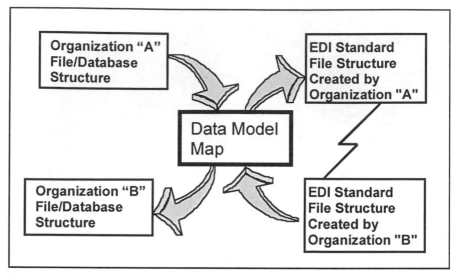

Figure 4.7 Using the Data Model as a Map

Knowledge Base

The next level we call a knowledge base (see Figure 4.4 column 3). Here all information about a product is integrated in the database. In addition to the functionality of the shared database, this level provides the ability for users to integrate their incremental learning 'knowledge' as they gain operational experience. As reliability and operational experience is gained it is related to the data, thus enabling users to make better business decisions. Decisions can be made based on self experience and facts (reliability statistics) derived from the past actions of others.

As an example, let us say a unit malfunctions in flight. The discrepancy is transmitted to a ground station and analyzed using databases that include aircraft maintenance information, flight scheduling criteria and troubleshooting guides. The system mines through the information and highlights possible problems or patterns and presents the knowledge, past actions and success rates for use in making business decisions. Now, the trouble shooting information tells us that the fault is either unit A or unit B, and our reliability knowledge base tells us that 60 per cent of the time unit 'A' is at fault and 40 per cent of the time unit 'B' is at fault. We can also be presented with the rectification elapsed time stating unit 'A' will take four hours to replace and unit 'B' will take 30 minutes to change. Station availability of the required parts and materials is presented indicating both units are available in stock at the base. The Minimum Equipment List (MEL) constraints are presented, indicating we are not permitted to dispatch with either unit inoperative or deactivated. Additionally, the scheduling criteria indicate we have one hour ground time prior to next departure. Based on these facts, the user can make an appropriate business decision. They can opt for the 60 per cent probability, notify operations and coordinate for a flight delay, or opt for the 40 per cent probability and potentially dispatch on time, always conscious of the risks and ramifications of their decision.

In the data interchange category we are now looking at not only supplier to operator but the addition of operator to supplier, operator to operator, operator to regulatory agency, regulatory agency to operator and supplier to supplier. We are looking at full product data, fully integrated. Information supplied once and only once. Everyone concerned uses that very same information, and utilizing text, graphics and voice, as a means of information retrieval.

Closing Thoughts

We have discussed the progression through different generations of information management and interchange. This has been a simplification of the actual operational environment where it is possible to utilize a mixture of technologies from various generations. Upon self reflection operators may find themselves, let alone them and their information suppliers, at varying points within the different cells (technologies) while also spanning different generations. Irrespective of where you stand today, hopefully this information provides an understanding of the strategic vision and direction all must take. Although current events in our industry may slow the pace of technological change, it has increased the awareness for secure information and quality data. Hopefully this will be seen as an excellent opportunity for all members of the aviation industry to enter into coopetition. 'Coopetition' is a neologism, defined as the act of working with others toward a common goal with the intention of competing against one another once the goal has been attained. Operators need to cooperate leveraging their combined knowledge and experience to establish the industry standards that will meet or exceed their common need and propel us into the Third Generation of standards.

Regardless of the information category or generation in which you are operating or striving to achieve, success is measured in terms of getting 'better'. For the purpose of this discussion 'better' is defined as cheaper, faster, more agile and flexible while maintaining (or improving) safety and security. Getting better in any information generation revolves around quality data (see Figure 4.8) which may be characterized as data which is without typing/spelling error, correctly tagged, conforming to the specifications for properties and domain and consistent across its usage. Quality data implies that modifications are made throughout. Partial modifications, where some data have been modified and other data have not, are not allowed. Quality data must have referential integrity. That is, any reference to or from other places or documents is valid. In addition, the data must be secure by ensuring that the data remain intact. The data must not be susceptible to intentional or unintentional alteration without proper authorization and approval. Finally, quality must be built in from its creation. We need to ensure quality data delivered to an agreed to information architecture supporting well-defined processes.

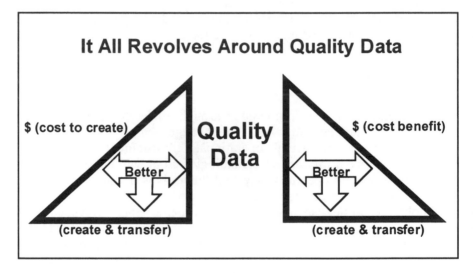

Figure 4.8   What is 'Better'?

References

ATA (2001a), *iSpec 2200: Information Standards for Aviation Maintenance* (CD-ROM), Washington, DC: Air Transport Association.

ATA (2001b), *ATA Interchange COO, ATA Information Interchange Concept of Operation*, Washington, DC: Air Transport Association.

Porter, M.E. and Millar, V.E. (1985), 'How Information Gives You Competitive Advantage', *Harvard Business Review*, 63:4, pp. 149-160.

Chapter 5

Flight Operations Information Interchange

Rick W. Travers

Lack of a Common Data Standard

Most in the aviation industry are surprised by the lack of standards in the interchange of flight operations information. They should be too. When we think of the aviation industry, we think of the latest and best innovations that today's technologies have to offer.

Background

There are numerous activities generating operational information: development of Flight Crew Operating Manuals (FCOMs), Advanced Qualification Programs (AQP), Crew Resource Management (CRM), Flight Operations Quality Assurance (FOQA), Line Operations Safety Audits (LOSA) and a myriad of applications of federal and civil aviation regulations (FARs, CARs, JARs). To interrelate and compile this information into a common context is almost impossible. In fact, the definitions used in these programs differ from source to source. In many instances, even the basic terms are used inconsistently. Disparities of commonly used language are exemplified when cross-referencing information from one program to another.

So what do the latest information technologies have to offer in aid of disseminating all this information? Like most industries, we have embraced the concept of electronic storage and delivery of information. Unlike most industries, flight operations have not embraced common terminology nor data standards. How can this be? Ask yourself, 'What is the common industry definition for information regarding the act of take-off?' Now ask yourself, 'What is the common industry definition for information regarding the aircraft landing gear?'

In the first instance, the definition of take-off will have much to do with the source in which you look it up; in the second, the definition is well established. Aviation maintenance/engineering personnel worldwide can easily find and cross reference information and related procedures catalogued by the Air Transport Association (ATA) System Specification (32) Landing Gear. As the industry develops electronic storage and retrieval systems, the management of information in maintenance/engineering is done rather seamlessly based upon the ATA System Specification. Not so in flight operations. Unlike the ATA Systems Specification established over forty years ago, flight operations have not established an information standard. For example, the use of the ATA System Specification, even at a very limited level, has not been widely promoted between maintenance/engineering and flight operations.

There are many examples of a lack of standards throughout flight operations information. However, one of the most striking is the lack of standards for phases of flight. Without industry standards, information managers are defining phases of flight to suit only their applied needs.

An overview of the requirements for phase of flight definitions reveals numerous specialized applications without consolidated cross-functional applications. In other words, the phase of flight definitions being applied under AQP training have been devised independently by corporate programs and training developers, not by an industry standard. Similarly, FOQA programs are developed and currently being approved without common standards between one corporate standard to another. The latest LOSA programs developing under the auspices of the human factors umbrella are now being devised with their own definitions of flight phases for analysis. Aircraft safety and accident and incident investigation efforts require phase of flight categorization and at the time of this writing are considering the use of a standard developed by the ATA. Meanwhile, every manufacturer and aircraft operator uses flight phases for structuring standard operating procedures (SOPs) and these are individually based upon corporate legacy or independent analyses.

Training issues are dealt with in an exhaustive manner at every level of flight operations, but these too reflect a varied application of phase of flight information standards. Each training organization, whether it be operator, manufacturer or third party vendor, independently develops training material for flight crews. Each of these groups may develop phase of flight definitions for their own use.

Standards are promoted, and most often mandated, within each corporate entity. In every part of the world, operators boast of internal documentation standards upheld to be crucial in the support of safety and SOPs. The need has been identified and encouraged (Degani and Wiener, 1994 and FAA, 1994). Some extensive systems are put in place, sometimes at great expense to the developer, and often grandfathered on corporate standards that have no relationship with an industry standard.

There are also inconsistent flight phase definitions in the use of flight warning devices (FWC, GPWS) and differing definitions for flight management computers (FMCs). These inconsistencies are at times used side by side in documentation for a single aircraft type.

From one operation (maintenance/engineering) to the other (flight operations) there is little common document standards today. The source of aircraft operational information is delivered in diverse means and structures, even from one manufacturer to another (see Table 5.1).

Current View

When considered in the context of information management at an industry level, the lack of standards becomes a long-term hindrance for the sharing and cross-reference of flight information. How can an industry that prides itself in the area of self-analysis for the benefit of sharing information produce data on a worldwide scale without basic standards? Once again the comparison to the ATA System Specification (based upon the ATA Spec 100) comes to mind. The aircraft system specification is not rewritten by every organization that has a need for aircraft part and component reference. Why should every organization with a need for flight operations reference rewrite a flight operation information standard? The quick answer to the question is that we have not had an industry standard.

Without data interchange standards for operational information, most major aircraft operators have done what they require to standardize documentation structures in their own organization. Every major operator is now conducting internal studies of their own corporate information structure for the future. Why now? The reason is that the electronic information technology has exploded. Electronic display of information has become the norm worldwide in every aspect of industry and commerce. This is a reality, and aviation is no exception. The production of paper manuals and the efforts in update and delivery of time-critical information has changed forever. Expectations for improvement in man-hour production times and the efforts applied in information delivery have been heightened everywhere. There is a flood of endeavors to fulfill these expectations.

Information technology experts not familiar with the aviation industry are surprised that there are no industry standards from which to start in developing proprietary or open source electronic standards. This is key in identifying the real lack of industry data standards. The need for an open source data standard is highlighted. Along with the need for this open source standard comes the need for an agency sanctioned to write, promote and uphold the standard.

Table 5.1   Manufacturers' Aircraft Operations Manuals Structures

	AIRBUS	BOEING	BOMBARDIER	EMBRAER
VOL 1	Systems Arranged by: ATA System Spec	Limitations Normal Procedures Supp Procedures	Systems Arranged by: Alphabetic* (Different titles than Boeing)	General Limitations Emerg/Abnormal Normal Proc Performance Flight Planning Weight and Balance Loading Config Dev List Min Equipt List Emerg Info Emergency Evac Grnd Servicing
VOL 2	Loading T.O. Perf Land Perform Special Ops Flight Planning	Systems Arranged by: Alphabetic (but with different titles than Bombardier)	Limitations Checklists Normal Ops Supp Procedures Emergency Abnormals Performance Spec Ops In Flight Checks	Systems Arranged by: Chronological Use
VOL 3 & 4	Limitations Abnormals Std Oper Proc Supp Technics In Flight Performance Engine Out Ops FMGS Volume			
QRH	Emergency Abnormals	All Non-normals All Performance	Warnings Cautions	Normals Section Abnormals Emergency Cautions
Format	SGML w/FrameMaker® CD-ROM (html) Word® Paper	Some SGML FrameMaker® PDF Word® Paper	Quicksilver® / (Interleaf®) Paper	Word®, PDF Paper
Page	5.83 x 8.27 (A5)	5.5 x 8.5 (Half Letter)	8.5 x 11 (Letter)	5.5 x 8.5 (Half Letter)

The idea of a flight operations information standard has been broached many times before. For a number of reasons it has not been established–the most prominent is the insistence of the end user in defining documentation structures for their own corporate use. This is not an approbation of blame, but a historical observation. It is true even today, and the very efforts to standardize data in support of the industry as a whole is wrought with this end user resistance to change at the delivery level.

The picture that is painted thus represents an area of the aviation information industry close to turmoil. If the proliferation of information is not harnessed within some widely accepted standard, flight operations information will remain as disjointed electronically displayed text on screen or paper The links, conditional processing, storage, retrieval and management of data by today's electronic standards, not even considered ten years ago, are now essential. The need for a widely held standard to harness information is desperately required in order to identify and harmonize operational data and exploit the power of the information technologies at hand.

Who will provide the industry with a flight operations information standard? Could a single manufacturer, given each with their own proprietary means of information delivery, agree to supply a common open source standard? What about promoting a standard written by one corporate sponsor (aircraft operator) or another? Could an industry agree to accept the work in an open source format of a single program (CRM, AQP, LOSΛ, or FOQA) and expect that program to maintain it? Could the industry agree to a third party proprietary solution implicating the possibility of perpetual marketing? Are international standards organizations interested and, if so, do they have industry and regulatory support? All these solutions are in play at the moment, but before a solution is found, the question of industry requirements must be addressed.

Display of Information Versus Management of Electronic Data

There are several proprietary methods for the electronic display of operational information available in today's market. Some of these solutions have near-term application, but they should not be confused with aviation industry standards.

Word Processing Text Displayed on Screen

In the rush to make information available electronically, a major step in the evolution of information management is being made. Flight operations information has historically been delivered as text on paper. First by wholly typewritten pages, next by photocopy machines and more recently by the use of word processors. Today, word processor text is being processed by more sophisticated document management software. It is this documentation that is being accessed on computer screens everywhere. References are being made to the new technology in order to display documents on screen. Inside industry observers are watching the transition of existing documentation to computer screens suspiciously. They should.

Information technology experts identify this limited form of electronic display as a near-term solution to a problem. In the short term, the technology easily allows conversion of legacy documentation to electronic display. This is not necessarily a bad thing; indeed it satisfies the desire for computer displayed information. For some smaller applications of documentation delivery, this should suffice. But what of the aviation industry information infrastructure as a whole and the flight operations data specifically?

The desire for electronically delivered operational information is wide spread. The display of text on screen can be promoted as electronic data interchange to the uninitiated who view electronic information as a viewer with word search capability and some basic subject linking. Electronic information delivery in this context is very limited, and its promotion as an industry solution to a data rich electronic environment is questionable. This limited technology has provided a simple means to display documentation on a screen. However, the linking capability of an industry information infrastructure is not supported by word processing text. What is missing is the identification of data standards in the evolution of flight operations information standards (Figure 5.1).

Identification of Data

Electronic data should not be confused with the word processing text currently being displayed electronically. In some instances markup language is applied to documentation, or the information is stored in database software. In these instances, the concept of data development is promoted; but promoted by whom and for what purpose? What industry standard is it based upon? Is the technology and the 'standard' to be applied in one program and not another, in one corporate entity and not another? Proprietary development of data technology solutions for an industry is certainly not in the best interest of the industry as a whole.

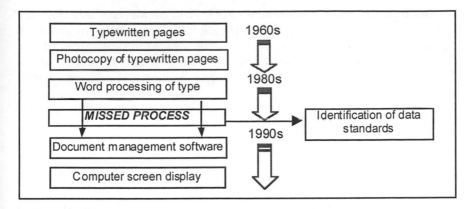

Figure 5.1 A Missed Process in Information Development

With advances in the management of electronic data, the limited standards used in flight operations information stand out as woefully inadequate. In every area, from suppliers, research groups, training development and support programs, the basic terms have not been harmonized. Harmonization and large scale agreement of terminology is required in order to implement an electronic data standard. Indeed, the use of common terms for even the basic systems on an aircraft has not been agreed upon within flight operations as has been done in maintenance/engineering. The terms used in 'phase of flight' have so many definitions that they cannot be used reliably to refer to a specific set of actions.

The standards used to manage information in some industries have been developing over many years. For example, the banking industry has established very sophisticated data standards out of the need for accurate communication between banks, their customers and other institutions allowing transactions worldwide. The ability to manage data for the betterment of an industry as a whole is heightened by the availability of recent data management technologies. Compare these with that of the flight operations information standards. Without commonly defined terms, information cannot be tagged accurately for electronic data interchange. Word searches and text viewers are all that can be hoped for in the future if a data standard is not agreed upon. Certainly flight operations deserves better than this.

In order to develop data for use with the current information technologies, an internal study of the requirements is mandated. How information is currently structured and delivered must be examined. Consistency and inconsistencies must be identified. Similar information under diverse hierarchies must be classified. A full understanding of the flight operations information taxonomy must be realized.

Additionally, the development of a data model to describe the information employed in flight operations is required. In order to define, merge and cross-reference with the content in other areas of the aviation industry, the data model must be compliant with related areas. The data model must provide basic linking of diverse data. Furthermore, the data model must facilitate automatic links to important information relevant to queries that may not be so basic or obvious.

Data modeling forces the application of common terminology (Dahan and Travers, 2000). It encourages users of information in the aviation industry to agree on an applied vocabulary. Inconsistencies can be addressed by either forcing applied definitions or by noting the contradictions and electronically mapping them. Documentation on a grand scale must be examined and the data model proven by decomposing current information into information objects. A robust linking system must be employed to ensure all information objects can be reassembled based upon a query for specific information.

The key to the functional use of data is the acceptance of data interchange standards. If data are to be used effectively throughout the industry, then a standard must be identified as to how this will be done. Again, this concept is not dictating how to structure information for internal use. It is an interchange standard for the efficient and automated exchange between organizations. How the data is managed internally is the choice of the corporate user or publisher.

Management of Data

Flight operations must divorce itself from the concept of managing just documents or text. Today's technologies support, and indeed require, the use of database content. Information restricted to page-based viewing principles has all but disappeared with the widespread use of computers and the Internet.

Documentation for different groups of end users should be developed from data. Change the data once it becomes effective wherever that data is required. The repurposing of source data is the basis by which information systems are built. A flight operations data standard will provide not only ease of changing linked information, it will also provide the means to discover new links. There are obscure links throughout current Flight Crew Operating Manuals and Flight Crew Training Manuals. These bits should be able to be identified and the source uncovered with properly conceived data standards.

In order to successfully manage data, additional information is required for each data object. This data about data, or metadata, may include the source or requirement for the information, and the date and effectivity of the data. Shared metadata standards are sometimes referred to as 'generic resources', a designator, because the same data resources can be shared amongst different users like maintenance, engineering, manufacturers, flight operations and training. If the standard for metadata can be used to support information for a variety of business functions in the same industry, repurposing is facilitated.

When metadata is shared through a common, open source, database standard, common use of software and hardware applications can be developed. The business of proprietary engineering and development is thus encouraged to provide the tools by which the data can most easily be managed within the industry. Indeed there are many and varied information technology firms ready to provide support functions to flight operations when such standards are provided.

Cross-functional Use of a Data Standard

Data can be managed in a manner required to deliver a consistent product. Moreover, the product can be updated consistently with the change of source data. Technical information related to specific airplanes does not change dramatically in time. However, when it does change (due to a change in the architecture or component configuration), the snowball effect can be dramatic.

A small change in modern aircraft software can have astounding ramifications. The effectivity information related to a change in software updating the related maintenance/engineering data should automatically revise and update the relevant flight operations data. This is a basic requirement that can be supported by data standards where a flight computer change on an aircraft done by maintenance/engineering links to the update of flight operating procedures of that aircraft. Shared metadata between flight operations and maintenance/engineering enables the revision and the effectivity of the information set for that airframe.

With an inquiry for information in one area of flight operations, the user should be notified about related areas. Indeed if the user is changing information in one area of business, the other areas using this information should be identified. This is a basic premise of business, and can be supported by common data standards. For example, a query regarding a standard operating procedure linked to the training information within the organization (and possibly the corporate standards).

Important information in one support program should be denoted in another. Not only to point to other related programs, but to induce cross-interest from one program to another. This appears fundamental, and can be supported by common data standards. For example, FOQA observations linked to HF, LOSA and flight safety data.

Subject areas have indispensable links to one another. Obvious as well as latent information can be assembled together. Aircraft systems, phases of flight and the environment can be intertwined resulting in conditional processing of information. This is a powerful tool for flight operations use, and can be supported by common data standards; e.g., procedural limitations in the use of flaps (system) during taxi-out, take-off, initial climb, approach, landing and taxi-in (flight phases) during a winter snowstorm (environment).

Future of Aviation Operational Information

The requirements for information interchange have never been greater in the aviation industry. To meet these requirements, the need for information interchange standards is very real. The basic means by which an industry communicates in the future is at stake and must be addressed. Already numerous organizations and interested parties are suggesting idiosyncratic solutions. Some can provide proprietary data management systems. Nonetheless, the industry as a whole desperately needs open source data standards in order to take full advantage of the technology at hand.

The Air Transport Association (ATA) is addressing many of these requirements. The lack of standards in operational information has been identified. The internal study of information structures, the data model, the generic metadata applications and the advancement of some basic standards are being submitted to the industry at large (Calderone and Travers, 1998). Manufacturers, operators, third party vendors and information technology experts are involved in this process. The goal is to provide an open source, flight operations data transfer standard. The business case has been identified and the work is ongoing. Manufacturers, operators and third party vendors are taking an active role in providing input in order to identify requirements and fulfill these needs. State and worldwide regulators are noticeably absent from these efforts, although most regulatory agencies dictate sweeping document standards.

The ATA has maintained Specification 100 (Aircraft System Spec, harmonized for electronic delivery in iSpec 2200) for many years. If there are changes to be made, updates or improvements, the ATA facilitates that process. This is important. Today, while there are a number of activities developing information structures for flight operations, none have an ongoing interest of the industry as a whole. While many interested groups define their own needs for structure and definition of information, there is little being done at the industry level, with the exception of the work under development by the ATA. Indeed the argument for the ATA to inherit the flight operations data standard is sound. The use and linking of information with the ATA System Specification, along with the sharing and promotion of common metadata applications, make the union of maintenance/engineering data with flight operations data a good fit. The ATA is logically the organization to develop, promote and arbitrate the flight operations data standard, as well as the System Specification. The ATA must find the resources to continue the development and promotion of these standards, and to promote cooperation between regulators, manufacturers and operators.

Industry Promotion

The development of a flight operations data transfer schema requires some agreement of definitions amongst users. Such agreement must be made to affect a data model discussed earlier. Fully developed data models need to have defined entities, elements and attributes. Each of these requires careful industry introspection. The work in this area was done by the Flight Operations Working Group (FOWG) under the auspices of the Technical Information Communications Committee (TICC) of the ATA.

The ATA already has a rather mature data model in which maintenance/engineering elements are defined. The Technical Information Communications Committee through the Business Interest Architects group (BIA) of the ATA has identified the links to the Flight Operations data model. (see Figure 5.2) Indeed the same definitions used for phases of flight are integrated through the applicable areas of maintenance and engineering.

The ATA Common Source Data Dictionary (CSDD) has been reviewed for inconsistencies regarding the implications of flight operations terminology. Further harmonization will be required in the future as the implications of the schema are refined and the processes of data mining and decomposition are advanced.

The models for the use of metadata are shared between flight operations and maintenance/engineering. Often referred to as 'generic resources', the metadata identified to be shared include the application of 'revision' and 'effectivity' functions.

The link of information sets has been clearly identified between ATA systems and ATA promoted phase of flight. Combined with environmental factors and aircraft state (normal, abnormal, flap and gear configurations), the decomposition of information is in process at the time of this writing.

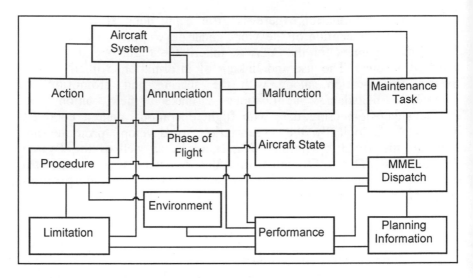

Figure 5.2 ATA Flight Operations Data Model (Simplified)

Levels of information importance have been identified. The principle of 'need to know' information is highlighted. The concept of information levels is advanced (Blomberg, Boy and Speyer, 2000) with background and support information available as needed.

The use of World Wide Web Consortium (W3C) open source information transfer protocols is the foundation for the information exchange. W3C standards can be applied to shared metadata with aircraft maintenance and engineering. It is in the interest of the aviation industry as a whole to support and indeed embrace the development of some widely applied information standards to further electronic data communication. The world's major aircraft manufacturers have identified the need for information standards for the next generation of aircraft. Major aircraft operators have identified the need for common data transfer technologies. Promotion of some standards today can target a wider application of an integrated system for the future.

ATA System Specification

The acceptance of the ATA System Specification at a two-digit level to identify aircraft systems information is an easy fit for flight operations. The Master Minimum Equipment List (MMEL), as used by most flight operations, is an accepted application of the standard. The standard is often used in reference to components, manufacture and maintenance of aircraft.

A more widely applied use of this established specification is supported for the development and identification of data in flight operations. Specific systems have been identified for use in flight operations (see Table 5.2). Others may be applied as required. The specification is mature and thus can be used in all areas. ATA supports, updates and replies to queries regarding the specification and can apply flight operations concerns to the industry now and in the future. The Joint Aviation Requirements (JAR-OPS) for Commercial Air Transportation endorses the use of the ATA Systems specification for flight operations information.

Table 5.2   ATA System Specification for Use by Flight Operations

20 General
21 Air Conditioning & Pressurization
22 Autoflight
23 Communications
24 Electrical
25 Equipment
26 Fire Protection
27 Flight Controls
28 Fuel
29 Hydraulics
30 Ice and Rain Protection
31 Indicating & Recording
32 Landing Gear
33 Lights
34 Navigation
35 Oxygen
36 Pneumatics
38 Water & Waste
45 On Board Maintenance
46 Information Systems
49 Auxiliary Power
52 Doors
56 Windows
71 Power Plant

ATA Phase of Flight Specification

With no accepted standard for phase of flight parameters in the industry, the ATA FOWG developed a widely applicable specification to meet industry needs (see Figure 5.3 and Travers, 1999). This is to become an 'approved for use specification' in 2002. This specification has already sustained proof of use as it has been applied in document standards by a major carrier and a manufacturer. It is currently promoted by safety audit groups, training organizations and by other organizations in need of operational information standards. The phase of flight definitions are tailored to commercial aviation flight crews. These definitions are broad enough to include other fixed-wing aircraft operations. Additionally, maintenance requirements and ground operations not related to flight crews were identified in the make-up of an integrated data model.

Human factors and cognitive task analysis techniques were applied in developing the phase of flight concepts (Travers, 2000). For example, the transition to 'initial climb phase' is marked by the crew beginning acceleration to a climb profile in the direction of the destination after complying with the procedure for noise abatement, obstacle clearance and/or complex Standard Instrument Departures (SIDs). Altitudes associated with these maneuvers are not fixed, in order to comply with the vagaries associated with any number of airports. The transition point is chosen by the crew and can be modeled by flight data in relation to that choice. Procedures may be applied to each phase as needed. For example, a hold procedure may be applied in the initial climb phase (for a gain in altitude in reduced airspace dimension) or in cruise, descent or approach phases. Non-normal or 'control configuration' procedures may be applied within any phase and indeed may require a shift in phase by the crew. The data required in certain phases might only be made available in a particular phase pertaining to non-normal procedures. Transitions from one phase of flight to another are rendered in the model. Logical transitions to reverse the procedural tasks are also depicted between phases in Figure 5.3.

As with the System Specification, the ATA will support, update and reply to queries regarding the Phase of Flight Specification ensuring that the standards can be applied to the industry now and in the future.

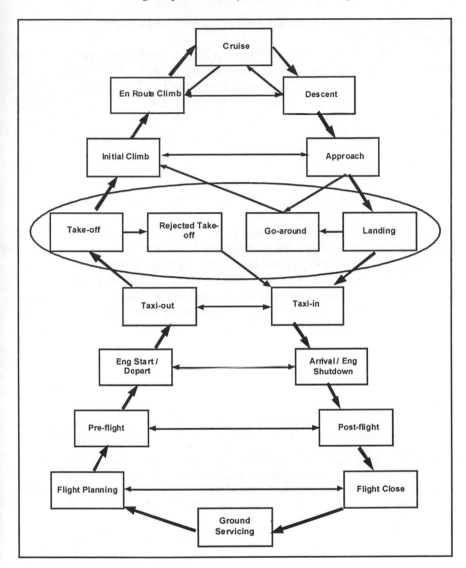

Figure 5.3 ATA Standard Phases of Flight

Summary

This chapter began with an explanation of the lack of real information standards in flight operations. An explanation has been given as to why the current paper-based paradigm of information transfer is no longer adequate to manage data. It is acknowledged that information, when available in an open source data schema, can be shared and repurposed amongst all industry users.

What is left to be done? The industry as a whole must put forth an effort to follow through with the initiatives of the ATA and others in support of common data interchange standards. The promotion and acceptance of even small initiatives with the use of ATA specifications will follow with even larger initiatives. The NASA/FAA Operating Documents Project (NASA/FAA, 2000) has endorsed the efforts of the ATA FOWG. The International Civil Aviation Organization (ICAO, 2001) has acknowledged the ATA Phase of Flight Specification. The International Airline Transport Association (IATA) will use safety audit data classifications with the use of the ATA Phase of Flight Specification in 2002. Indeed the Air Transport Association requires the support of the industry in order to continue its work in the area of standards and the applications of supported specifications.

Flight operations information users and regulators must acknowledge that there is a target standard and begin to understand that initial steps can be made. Further promotion will result in wider aviation data standards for the industry and a betterment of the information dissemination for all users.

References

Blomberg, R.D., Boy, G.A., and Speyer, J.J. (2000), 'Information Needs for Flight Operations: Human-Centered Structured of Flight Operations Knowledge', pp. 45-50, *Proceedings of International Conference on Human Computer Interaction in Aeronautics*, Toulouse France, Cepadues Editions.

Calderon, V. and Travers R.W. (1998), *Flight Operations Technical Information Interchange Master Document*, Washington, DC: Air Transport Association.

Dahan, B. and Travers, R.W. (2000), *A Data Model for Flight Operations Information*, Submission to The European Institute of Cognitive Science and Engineering (EURISCO).

Degani, A. and Wiener, E.L. (1994), *On the Design of Flight-Deck Procedures*, (NASA Contractor Report 177642), Moffett Field, CA: NASA Ames Research Center.

FAA (1994), *Air Transportation Operations Inspectors Handbook*, 8400.10, Washington, DC: Federal Aviation Administration.

ICAO (2001), *ICAO Annex 6, Proposed Amendment to Operation of Aircraft Part 1*, APPENDIX 3, Operator's Flight Safety Documents System, 4. Standardization of phases of flight, International Civil Aviation Organization.

NASA/FAA (2000), *Developing Operating Document: A Manual of Guidelines*, Moffett Field, CA: NASA Ames Research Center.

Travers, R.W. (1999), 'Phase of Flight Standardization', *Proceedings of the ATA FOWG*, Sept. 13-16, 1999, Washington, DC: Air Transport Association.

Travers, R.W. (2000), 'Pilot Centered Phase of Flight Standardization', pp. 51-56, *Proceedings of International Conference on Human Computer Interaction in Aeronautics*, Toulouse France, Cepadues Editions.

Part 2
Management of Aviation Operational Information

Chapter 6

From Documents to an Aviation Information Database

Terry J. Snyder and Anita Kanakis

Introduction

Advances in publishing technologies have unshackled information that has been imprisoned within proprietary publishing application documents. 'Sharing' information is no longer limited to cutting and pasting blocks of text from one document to another, often requiring an additional conversion process. Publishing is moving toward 'sharing' in the truest form of the word - actually reusing content stored as objects within a database. Instead of mere characters grouped as words, the text is actually 'intelligent' - that is, it 'knows' and identifies to the system what topic is covered within it. It also 'knows' what kind of text block it is (body, title, list, etc.) and its appropriate relationship with surrounding text blocks, thanks to document and content structuring. This chapter focuses on transforming imprisoned, isolated content into portable, contextual building blocks of optimized data and the subsequent management of the content object repository database.

Portability, intelligence and quality increase the value of the data tremendously to both author and customer. However, this 'enhanced' data comes with increased and more widespread responsibilities. If data are inaccurate or mislabeled, sharing means a much wider and quicker dissemination of wrong information, rather than containment within one document. The potential negative impact on operations and, more importantly, safety is obvious. Thus, this chapter also provides workflow and publishing process concepts to ensure working groups create and manage shared data optimally.

Structured Information and Data Management

What is Structured Data and Why is it Necessary?

To reiterate, the ability to effectively manage and reuse data in a timely manner among different publishing media depends on the data being structured correctly according to a standard rules-based markup language. Defining the logical structure and then using a markup language such as SGML and XML brings intelligence and portability to the data. This allows for delivery of information through different media types and formatting engines.

Structure defines what the data is, not what the data looks like (St. Laurent, 1999). A caution within a flight operations document has a different function than a standard paragraph, and its definition and controlled availability in the structure reflects that difference. Structure also establishes the critical 'parent-child' relationships between text blocks to ensure the data are used in the proper context. How those blocks of text visually appear, are formatted, is determined by the requirements of the end user (audience) for each medium the data is released in. This division between structure and format requires two different kinds of data 'ownership' - central content ownership and (usually) multiple release medium ownership. Processes for publishing structured data must support this dual ownership and accountability. Specifically, structuring information provides the benefits listed in Table 6.1.

How Structure in Documents Supports Data Management

Without structure and intelligence, content within documents must be read, analyzed and verified as appropriate for reuse in another document by a human resource alone. Significant manual research must be accomplished to ensure the content is current, accurate, approved and in the appropriate context for reusing in another document.

Content within structured, intelligent data objects, on the other hand, has a virtual history dossier transparently supporting it, details formerly possible only at the document file level. This subfile identifies such attributes as ownership, authorship, currency, version, sources and approvals, in addition to any useful hyperlinks to other pertinent content.

More importantly in a world of data sharing, each data object is defined in relational context with surrounding content. Each object has its place in a structured 'family tree', identified as a subset or 'child' to a larger 'parent' topic, and perhaps as a parent itself to subsequent child objects. When an object is imported into another document outside its source document, structure ensures all key data related to the object may at least be considered for accompanying the imported object in order to prevent a loss of context.

As would be expected, the more details available about the nature of the data, the easier it is to manage for multiple use.

Table 6.1   Benefits of Structuring Information

Information Management Benefits	Identifies what the data is and its relationships to other data.
	Ensures blocks of content are reusable as single-source multi-use objects.
	Ensures information is easily exchangeable between media and users.
	Allows data to be stored within an object repository, not imprisoned within a document.
	Separates the format from the data.
	Helps protect documents/data from being adversely impacted by stakeholders with varying publishing/formatting skills.
End User Benefits	Clarity enables document to more efficiently accomplish its mission.
	Intuitiveness and logical construction eases navigation through data.
	Patterns of information organization within levels of hierarchy establish data relationships.
	Consistent application of standard conventions and rules prevents misinterpretation of data.

The Value of Reusable Objects and Interchangeable Data

The concept of data stored as single-source multi-use objects is as valuable for the problems it prevents as much as the advantages it provides. In an unstructured environment, ownership of factual information is often attributed to the owner of a particular document. If the information appears in more than one document, it may be claimed by more than one owner. This may make it difficult to ensure all occurrences of the information are kept up to date concurrently, and to determine who has final approval when changes to the information are proposed.

Single-source objects have the advantage of having just one owner take responsibility for keeping the information up-to-date and accurate. Multi-use means the information is readily available for publishing in any medium, and is consistent from medium to medium should the end user access it in more than one location.

Possessing the ability to distribute the same text content in different media is not enough. Portability (interchangeability) also includes retaining the descriptive meta-information about the data, giving it the independence of application and platform. Traditional publishing tools support conversion of content in generic ASCII text blocks, with no retention of formatting, which is not in the best interest of the end user. Using resources (either qualified or not so qualified) to re-analyze and re-apply formatting to imported data with the risk of missing important parent-child relationships and emphases is undesirable and time-consuming.

Adding structure to document content using a markup language allows more universal access to the clear intent and utility of carefully crafted information. SGML and XML are widely accepted standards by many companies for ensuring their publications are rules-based, as the term 'markup language' implies. The interchangeability of rules-based data allows them to maintain less redundant information as well as realize lower costs associated with easy conversion from one application to another and fewer people needed to manage the data.

Moving Past Proprietary Documents

Converting content imprisoned within a proprietary document into structured data objects available for reuse simplifies the entire publishing process. One subject matter expert for the content of each topic, as well as one for each of the other tasks associated with preparing the content for publication, are identified and given the responsibility to carry out the publishing process. A major paradigm shift occurs as the efficiencies of managing single-source objects rather than maintaining parallel staffs to manage duplicated but isolated content in separate documents and publishing tools become apparent. The concept of structuring data is still relatively unapplied in the world of corporate data, as indicated by Figure 6.1.

Ensuring Professional and Consistent Formatting

Even the most coherent and well-written content can end up confusing and hard to use if arbitrarily formatted by a stakeholder without professional design skills. Both the final presentation (including the document's ability to smoothly map out in any medium) and the efficiency in producing the formatting can be adversely affected when left to non-qualified stakeholders. A professionally structured document protects it from the whims and varying publishing skills of stakeholders within the publishing process. Put another way:

> Authors are subject matter experts for whom it is a waste of time to be concerned with typography or document layout. Some call it creativity, but inside organizations where the author has free reign over format, the result is usually an incoherent collection of badly looking documents (van Herwijnen, 1991, p.6).

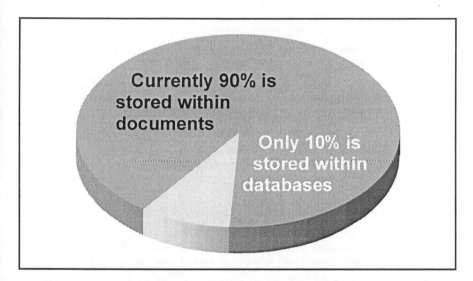

Figure 6.1   Corporate Data Storage Percentages

As authors and content contributors become increasingly aware of the basic rules and structure hierarchy of the documents they deal with, their jobs become increasingly easier. Reduced is the time spent determining layout, conventions, terminology and typography - this becomes an activity in simply selecting the appropriate publishing tools (template, style sheet, editing manuals) from approved and validated options.

The end users of the documents are even bigger beneficiaries. They find the information in the newly structured and standardized documents increasingly easier to navigate and retain. Content is logically constructed and presented to tap into an intuitiveness developed by a coordinated effort to provide the same basic rules to all the published material they access. These repeating patterns and levels of hierarchy establishing data relationships allow the reader to concentrate on the content without questioning variances in presentation and emphasis.

Steps in Optimizing Information for Efficient Management

Building the Publishing Infrastructure

To take full advantage of managing information that has been structured, the right resources and publishing business process must be in place. Most importantly:

- Establish the project management methodology and implementation plan.
- Identify appropriate team made up of expertise in the areas of:
    - Content subject matter (aviators, instructors, engineers)
    - SGML/XML (IT analysts, diagrammers)
    - Business process (stakeholder task team, efficiency experts)
    - Publishing standards (technical writers, editors)
    - Formatting (aviation document designers)
    - Information technology and architecture (programmers, analysts).
- Identify internal or external expertise for the conversion from static documents to structured documents.
- Develop an automated publishing business process and workflow.
- Provide workstations with structure-based authoring applications.
- Develop training and technical support resources for stakeholders.
- Ensure stakeholders and management commitment.

The establishment of a publishing infrastructure requires a wide range of resources. First, apply a project management methodology that ensures the project is properly recorded and managed. A team of subject matter experts for all aspects of the effort must be assembled and coordinated. Expertise (whether internal or external) must be acquired for the conversion from static to structured documents. To ensure that an adequate number of qualified stakeholders (by job description) are responsible for accomplishing necessary and well-defined publishing tasks, an automated publishing business process and workflow tool must be developed. A stakeholder matrix, as shown in Table 6.2, can help identify requirements by data topics and tasks. Required resources also include hardware such as workstations equipped with structure-based authoring applications and servers for the object repository and automated workflow, as well as training and technical support for stakeholders. Finally, it is critical that publishing stakeholders and management are committed to realizing the benefits of presentation and process standardization.

Table 6.2 Stakeholder Matrix

Data Topics	Data Ownership Tasks								
	Task A	Task B	Task C	Task D	Task E	Task F	Task G	Task H	Task I
1 **Exterior** **Inspection**	FTS	FTS	FTS	FMO	FTS	FTS	FMO	FTW	FTW
	PSC	PSC	OEG		PSC	PSC	FMA	PSC	PSC
	FTW	FTW	FAA		FTW	FTW			ART
	FMO	FMO			FMO	FMO			MED
	FMT	FMT			FMT	FMT			
		ART							
		MED							
		OEG							
2 **Cockpit** **Preflight**	FTS	FTS	FTS	FMO	FTS	FTS	FMO	FTW	FTW
	PSC	PSC	OEG		PSC	PSC	FMA	PSC	PSC
	FTW	FTW	FAA		FTW	FTW			ART
	FMO	FMO			FMO	FMO			MED
	FMT	FMT			FMT	FMT			
		ART							
		MED							
		OEG							

Converting Static Information into Structured Data Objects

The conversion process starts with several planning steps that may be overlooked in the rush to transform the information to data. The conversion process requires IT and publishing standards expertise up front, working with content stakeholders to analyze the documents. This analysis identifies the levels of document organization and necessary text elements based on end-user needs (see Figure 6.2). Instances of the structure are built and tested against data within the document to ensure validity. As development activities continue with the structure, a formatting expert (aviation document designer) should ensure the document design is consistent with company and industry standards (for industry standards, see ATA, 2001). Formatting based on a basic template for the category of document is tailored for and validated by specific end-user groups, such as flight crewmembers or training instructors. Each user-group template is based on common standards maintained in a master style manual. This activity may include building the style sheets or format specification for various output media types.

Once the structure and the format are approved, the actual conversion of the document can take place. Depending on internal resources and expertise, an outside vendor may be the best path to take. The conversion of a manual from a static environment to a structured one entails adding metadata, or data about data, to the data objects allowing for the identification of elements such as the source, history and applicability of the object.

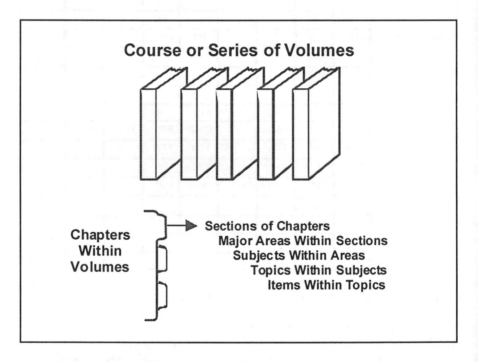

Figure 6.2 Document Organization Level Hierarchy

Document stakeholders must be trained at the appropriate time to accomplish tasks associated with publishing structured documents. Managing changes to the document is accomplished within the standard publishing business process and an automated workflow process customized to organizational requirements. Once the document is converted, the data is stored as a compilation of data objects on a database server. Finally, the specifications for both the structure and format must be carefully recorded and managed throughout the lifecycle of the documents they support.

Employing an Effective Process for Managing Data Object-Based Information

User-friendly processes and tools encourage all stakeholders to utilize the automation to achieve efficiencies in task completion, standardization and quality control. An effective process ensures the following:

- Interfacing with the process is as intuitive and smooth as possible.
- Items accepted for processing are valid for stakeholders to spend time on.
- The required task descriptions and which stakeholder is responsible for accomplishing each task are stated clearly.
- Process activities go to stakeholders' backups if stakeholders on prolonged absence.
- Priority and scope of activities are identified early on to enable stakeholders to manage their personal workloads.
- Data cannot be stored as an approved object available for sharing until it is signed off as valid, viable, accurate and complies with standards.
- The effectiveness of the process, tools and standards are measurable and allow for continuous improvement.

Stakeholders comfortable in a process and committed to its success are more likely to be interested in its continuous improvement and take the time to offer suggestions. The process administrator should make every effort to encourage feedback and act on suggestions in a timely manner.

Establishing Working Groups to Manage Shared Data

Data-Based Versus Publication-Based

Unstructured information in siloed publications often requires parallel stakeholders. Information that appears in more than one publication can be processed, sometimes quite differently, by several stakeholders with the same responsibilities but in different publishing organizations. Some stakeholders and tasks will be different based on the end use of the publication. However, there can be problems when more than one person is identified as being responsible for final content and release of the information. Examples of problems include confusion between versions or uncoordinated distribution of the information resulting in conflicts between revised and unrevised publications.

On the other hand, when the ultimate owner of a data object is clearly designated within a process which controls any release, end users can be assured that the content they are accessing is the same quality-checked content in every publication or program it appears.

Lessons of the Assembly Line

If everyone in the process does one thing and does it well, the end product usually reflects it. Less multi-tasking offers the opportunity and expertise to identify inaccuracies and do each particular task optimally and systematically. Show-stopping issues can be identified sooner to prevent wasted effort on the part of all stakeholders in the process 'line'. Fewer overlapping and/or multi-tasking activities translate into a more manageable workload and more accurate timeline estimates. Working groups in a structured world are more loosely organized to serve as best-practices validation groups rather than organized solely to support certain documents. They can be electronically linked rather than co-located or organizationally linked. One could draw parallels to a floor supervisor and quality circle group.

Simply, Rules Should Rule

A coordinated effort to publish consistently does not mean a complicated workflow. It does mean a commitment to publish according to rules that make sense to everyone in the working group:

- Standard operating procedures developed by publishing experts and validated by stakeholders.
- Communication and accessibility of approved standards.
- An official process administrator.
- Stakeholder and management commitment to the rules.
- An accountable owner of the standards and arbiter of issue resolution.

A commitment by all stakeholders to faithfully follow the standard publishing and issue resolution processes is necessary. A well-designed process stakeholders find easy to commit to rather than work around offers the best chance of consistently publishing quality products. Support of a workflow that helps stakeholders manage their workload and makes them look good at their jobs encourages continuous improvement suggestions. However, even a brilliant process can be ineffective if inadequately communicated and guidelines are difficult to access. Training, technical support and skillful editors are critical to success. Finally, the value of an entity officially accountable for administration of the process and stakeholder support cannot be overemphasized.

With all stakeholders publishing according to the same rules, customers benefit from information that is consistent and intuitive no matter what the medium. Having a clear roadmap also streamlines the authoring, proofing and editing activities and avoids time-consuming tasks involving comparisons, validation, researching for precedents and making style decisions. Workgroups are able to coordinate efforts and devise publishing issue solutions using a common publishing language and design philosophy basis. Collaborative efforts of qualified experts to develop and quality check shared data result in a better final product for the customer.

In Conclusion

The efforts to convert simple text into intelligent data, build the infrastructure to effectively manage it and faithfully follow quality publishing SOPs are challenging. One way to meet this challenge is to decompose the process into manageable steps that can be implemented in phases that meet near-term corporate needs while keeping on course with an overall vision. Savvy aviation companies recognize that trying to maintain a successful operation (against 21st century expectations and competitors) without making the effort to manage their data is even more of a challenge.

References

ATA (2001), *iSpec 2200: Information Standards for Aviation Maintenance* (CD-ROM), Washington, DC: Air Transport Association.
St. Laurent, S. (1999), *XML: A Primer*, Second Edition, Foster-City, CA: M&T Books.
van Herwijnen, E. (1991), *Practical SGML*, Netherlands: Kluwer Academic Publishers.

Chapter 7

Structured Information for the Cockpit

William W. LeRoy

Introduction

The flight bag carried today by commercial pilots, in conjunction with onboard aircraft systems, contains information needed to safely operate the aircraft in any flight environment. The information is logically and functionally organized into a series of operating documents called manuals, handbooks or checklists. Of these, the most common operating documents include a company policy and procedures manual (Flight Operations Manual), an aircraft operating manual (Pilot Handbook) and one or more chart and navigation manuals (Route Manuals). The structure is optimized for paper-based publishing and is constrained by size, weight and revision processes. As carriers move from paper-based information systems to electronic documents for flight and learning environments, there must be a significant shift from current content and organizational paper-based paradigms to dynamic, self-assembling information.

Paper-based information systems require a document structure, which is unsuitable for the cockpit, and moving legacy information to the same structure in an electronic format does not address the major issues affecting safety, efficiency and compliance. Due to size constraints of the flight bag and in the cockpit, paper-based information is not organized to meet information requirements, but rather to maximize the amount of information for the given space allocation. Due to both size and revision concerns, information is not repeated in multiple volumes, even when information in one volume requires information in a second to meet information needs.

This leads to another difficulty with the current paper-based system. Information needed for a specific flight environment is likely to be spread across separate physical volumes. This requires the pilot to extract information from multiple locations in one or more manuals and cognitively reassemble them into usable information. For example, at one major operator, a de/anti-icing event requires information from three separate manuals: the Flight Operations Manual, Pilot Handbook and a Route Manual. None of the manuals independently contain sufficient information.

Even with detailed indexes and other paper aids such as quick reference cards or books, pilots must still rely on their knowledge of where information exists, or trust memory for accurate information. This can lead to safety, operational and compliance problems.

This paper paradigm should not transition to electronic media. Even with electronic enhancements to provide rapid navigation, electronic documents structured around legacy paper volumes still require the pilot to locate and cognitively reassemble the information objects, in context, in order to have sufficient information to perform a required task at a point in time in a specific environment.

Legacy 'flight bag' information must be decomposed into specific information objects that can be recombined or altered based on probable overlapping events. This will allow application builders to create self-assembling information instances from stored information objects. To access information, the pilot would identify his or her location within an environment, and the system would return an electronic document containing the information, prioritized and layered, required for the event, and modified for additive conditions.

The shift to self-assembling information requires operators to 'decompose' current information structures into information objects and develop methods of combining these objects into larger objects. These objects, structured and in context, will supply the required information at a specific point in time in a specific environment.

Information Characteristics

Pilots require information based on their particular location within a flight environment. A flight environment consists of event sets; each set containing events marked by a beginning and end, and requires one or more tasks to recognize, confirm, control or correct. These events may be singular, simultaneous, or overlapping and modified with additive conditions. Industry-standard phase-of-flight definitions are available for transport aircraft, and can be used to define information requirements for specific events within those phases (Travers, 2000). Use of an industry standard further promotes a consistent information exchange between manufacturers, other suppliers and operators and reduces the development time for producing a useable information architecture. Information requirements exist at each location in the environment, at a specific point in time, as shown in Figure 7.1.

Information Components

Operational information uses three basic components that, when combined with context and structure, shape information. These components are data, concepts and instructions. They are the building blocks of the philosophies, policies, procedures and practices as proposed by Degani and Wiener (1994) along with supplemental information that provides environmental knowledge. For instance, supplemental information, such as how the auxiliary power unit works, cannot be properly categorized as one of the 4Ps, but it is essential to understanding a specific philosophy, policy, procedure or practice. Thus, operational information consists of the 4Ps plus supplemental information, and when the data, concepts and instructions are properly structured, it forms a construct that is understandable to end users, such as commercial pilots.

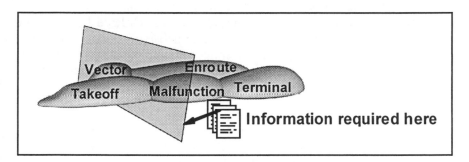

Figure 7.1 Information Requirement

Data consists of objects quantified by physical properties. These quantifiers include numbers, colors, physical objects, locations and settings. Examples include 'Flaps 20', FL400, red-guarded switch or Captain Winkler. Data provides little or no information by itself and must be linked by concepts and context before it can become useful as an information object.

Concepts are cognitive constructs required to understand how objects behave or relate in an environment. They provide a framework for processing these objects in order to identify one's position in an environment, and predict or facilitate a change to that environment.

Example: The concept of a stabilized approach is required for a pilot to determine whether the flight parameters (data objects) of airspeed, sink rate, aircraft configuration and airspeed will yield a satisfactory environment for successful landing.

Concepts may be layered or nested. This is important in building complex information objects, such as automation procedures. Stabilized approach may be used as a simple example of a nested concept.

Example: The concept of the stabilized approach requires the pilot to understand the relationship between airspeed, distance remaining and altitude change, along with the resulting, or nested, descent momentum, engine response capabilities and human factors considerations in recognizing and responding to specific hazardous situations.

Instructions are lists of actions needed to complete a task to confirm a particular environment or to affect that environment. For example, a checklist to troubleshoot an aircraft system to identify a faulty component identifies a location in a flight environment which now includes an additive condition in the form of an aircraft systems degradation or failure, and allows the pilot to affect a change to that environment either by a corrective or adaptive action.

Instructions include ordered and unordered lists. Ordered lists may be either procedures or flows. A procedure is a list of instructions that must be carried out in a sequential manner to complete a task; flows are user-developed lists to accomplish items required for a specific task. Procedures may consist of flows, but flows may not contain procedures.

Example: An 'Originating Checklist' for the Fokker F100 aircraft contains items that require both captain and first officer to accomplish certain tasks using a flow. The checklist item confirms a task (e.g.,. 'Panel.......Set'), which requires specific settings, but not an order for verifying those settings. Carriers often train specific sequences for a flow; however the pilot must only demonstrate that his or her flow is reliable and repeatable.

At a minimum, an information object contains one or more information components (data, concepts or instructions), bound by context and structure to provide the pilot information usable at the point in time in the flight environment. The smallest usable information objects are paragraphs, table cells, bullets and images.

Example: A special bullet, called a checklist item, may be a complete object, assuming the pilot is current and qualified in the aircraft, and is currently executing the 'Before Engine Start' checklist. He or she has complete understanding of the object, and requires no further information (see Figure 7.2).

APU.......................................Established

Figure 7.2 Minimum Complete Information Object

However, in most cases these objects are components of larger objects used to create complete information objects. Context and structure provide the essential connectivity between these objects to render them useful. Continuing the example above.

Example: If the pilot needs additional information to accomplish this item (e.g., in training), assuming a base level of knowledge, he or she might need the expanded instructions as shown in bold (see Figure 7.3).

APU...................................Established
    To start APU
-   **If AC power is not available**
    -   **Ensure BATTERIES switch is on**
    -   **Keep APU start selector......**

Figure 7.3 Multiple Information Objects

Context is the mechanism that provides information by relating one or more objects to other objects within the environment. For example, the data object '8,000 feet' does not provide adequate information for a pilot to pinpoint his or her location in an environment unless it is in the context of an airspace assignment (e.g., initial approach fix altitude) or aircraft operational limit (e.g., maximum cabin altitude). Without context, this data object has little or no meaning.

In most cases, an information object must be viewed within the context of parent, sibling and/or child objects. For example, when the information required is a list item (e.g., bullet), which is part of a 'Level 3' paragraph, the pilot must consider the particular list item in the context of other list items and the enclosing parent in order to create a usable information object. In short, if your answer is in the list, read the enclosing paragraph; if it is in the paragraph, read the enclosing section. This concept is important when considering usable information objects for specific requirements.

Structure provides a standardized, predictable means of organizing information objects in context. It allows the author to build successively more complex information objects by combining and organizing information objects to meet requirements in the target environment. Consistent structure allows the user to extract information efficiently by presenting objects in a predictable manner.

Example: Inconsistent use of heading levels requires the reader to reorder topics to understand the information.

In addition, structure is a critical means of supplying object context. For example, one structural method to facilitate object context is 'titling'. Titles for parent, sibling and child document elements often give clues to the context of the information object. Tables of content or bookmarks (Adobe PDF) give contextual clues to information by surrounding the target objects with similar or related objects.

A structural element that does not appear to the reader is metadata. Metadata, as shown in Figure 7.4, is additional description attached to objects, which provides the document builder a way to add information to information objects. It is a critical element in tracking information such as author, approval authority/date, ATA code and general document information, but also allows detailed description of information, including key words, phase of flight and event code (e.g., TPO/SPO), in order to manipulate databases and produce specific data sets.

Information Objects

Information objects have different complexity and characteristics, depending on the point in time in the environment; they divide into primary, expanded and supplemental objects.

The primary level is the minimum-sized information object required for the tasks, and assumes the pilot has sufficient skills, experience, situational awareness and task loading, to require a minimal object.

Example: A checklist item may be the primary object in the context of the associated aircraft and checklist. This item is usually sufficient given target audience and task (see Figure 7.2).

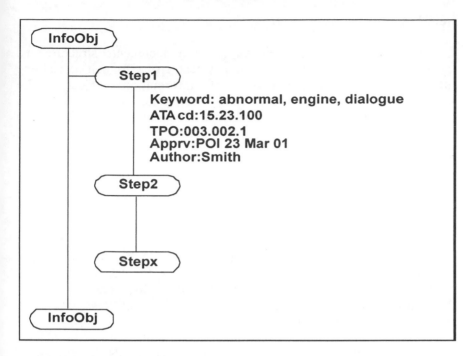

Figure 7.4　Metadata Example

The expanded level combines information to include both primary and expanded information. This is common on checklists items, as seen in Figure 7.3, as well as procedural objects such as dialogue boxes.

Example: Continuing with the example above, if the pilot is inexperienced or non-current, the expanded checklist item may be necessary to provide necessary information for the pilot to complete his or her task, as shown in Figure 7.3.

A supplemental object adds environmental information to both primary and expanded objects. This is used mainly in a training environment as the pilot establishes sufficient fundamental knowledge to require only primary and/or expanded objects. Once the pilot attains a level of proficiency, reference to most supplemental objects is infrequent.

Example: In the example shown in Figure 7.2 and expanded in Figure 7.3, a supplemental object might consist of company policy and procedure regarding when and how to use checklists.

Self-Assembling Information

Current systems will shift from paper-based documents structured to maximize information in a limited physical space to self-assembling information instances composed of information objects. The significant benefit of self-assembling information is the ability to collect and organize information objects into appropriate information instances.

> *Example*: At system initialization, the pilot inputs the aircraft tail number into the system, and later identifies a location (e.g., top-of-descent, Roanoke, Virginia). Based on all events within the event set, along with additive conditions, the system returns an information instance.

Self-assembling information has five primary characteristics. It considers the entire environment (e.g., normal operations modified for an abnormal condition); provides information for the current task; creates successive, prioritized information for subsequent tasks; links this information together; and, finally, establishes links to supplemental information.

A self-assembling information instance is built from data objects stored in a database. The instance is based on the pilot's description of his or her location in the environment.

> *Example*: Continuing the example above, the system assembles an information instance that includes a 'Preliminary Landing Checklist', 'Approach Briefing', standard arrival, terminal plate, 'Landing Checklist' and airport diagram (see Figure 7.5). The terminal approach plate would be customizable based on pilot preferences (e.g., view terrain or obstacles), and contain information for a specific aircraft type. In addition, a series of progressive links will take the user to different levels, including supplemental information.

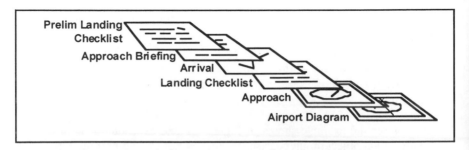

Figure 7.5   Top-of-Descent for the Cockpit Environment

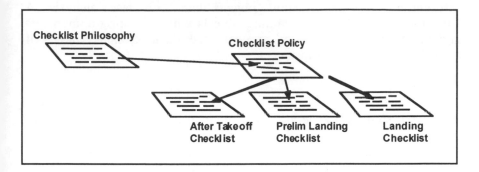

Figure 7.6   Top-of-Descent Learning Environment

These same objects could be repurposed for a learning environment and become elements of an instance that includes objects containing philosophy and policy as initial information objects, along with any required sibling or child objects (see Figure 7.6).

For some classic requirements, such as Federal Aviation Administration (FAA) approval or preparatory ground training, self-assembling documents will mimic their paper-based counterparts. However, the real utility of the system is the ability to repurpose information objects, based on user needs, to build mini manuals of information instances to meet specific requirements at specific points in time in specific environments.

An electronic document system based on self-assembling information takes input from the pilot and returns an electronic document consisting of prioritized, layered, linked objects, useful for the point in time at a location in a specific environment. The objects contain information for current and future related events. The major benefits are twofold. First, the pilot has complete information for the event(s) at hand and does not need to locate information in a multi-volume document. Driving down time to access information and providing it in a useful structure reduces task loading and leads to greater efficiency and increased safety margins. In addition, it enhances compliance, as information is no longer 'hidden' in structures designed to maximize the amount of information for a constrained physical space.

Conversion and Application

The information conversion process requires five important steps: 1) Developing a concept of operations regarding information, 2) Constructing or accessing a task analysis, 3) Decomposing current information into information objects based on task analysis, 4) Constructing a database containing objects with appropriate metadata, and 5) Developing a system to query the database and return self-assembling information instances.

Process Steps

This first step is to define what your information system should provide. It must consider the targeted audience, level of complexity (e.g., how many simultaneous abnormals), success criteria, capital requirements, return-on-investment, to name a few. A well-defined concept of operations will be a yardstick for nearly every decision in the development process.

The conversion process continues by defining flight tasks through a detailed task analysis. This is available for fleets participating in the Advanced Qualification Program (AQP), and is required to determine information requirements for all normal tasks as well as probable additional conditions. In addition to the design and population of the task analysis, organizations must invoke stringent policies regarding the processes and approval mechanisms to add, delete or change information. This is an additional paradigm shift for many organizations where aircraft fleets enjoy some level of autonomy and may be reluctant to accede to stringent process controls.

This process requires collecting all information relevant to each task. It should include, where appropriate, philosophy, policy, procedure, practices and supplemental information. The end result is the ability to query a database of objects and organize them for presentation based on environment. For example, a flight task generally calls initially for a procedure or practice, with philosophy and policy included along side supplemental information. An event in a learning environment would normally require a policy and/or philosophy element as an initial object. When possible, additive conditions requiring different elements should be included and identified through appropriate metadata. This is to allow the system to parse the objects and assemble information that considers the additional condition.

> *Example*: An engine-out condition might affect the type of missed approach at a 'Special Airport', and the system would identify the engine-out missed approach in lieu of a normal missed approach when constructing appropriate chart and navigation information.

Structuring these objects requires a markup language. The markup language must be able to enforce object structure rules, as well as associate metadata with a particular object. Publishing software like Microsoft Works or Adobe FrameMaker provides markup languages that can identify information elements. However, they do not have a robust ability to enforce structure rules, nor attach metadata. The two markup languages able to enforce structure are Standard Generalized Markup Language (SGML) and its offspring, Extensible Markup Language (XML). SGML has been in use for over a decade, and is a powerful tool used primarily for security of and repurposing information in a paper environment. Data structured via XML markup can be viewed by Internet browsers in current use, and can also be stored in and retrieved from databases.

This critical step must occur in concert with information decomposition. Understanding the desired product (concept of operations) is crucial to defining not only field, records and related information, but also object size and complexity and the method to attach descriptive metadata. In addition, the database must be capable of a high degree of interactivity with the user, as well as the display mechanism.

Finally, a unifying system must pass queries to the database and interpret and display the resultant data sets. This system must bind together diverse elements such as information, performance calculations and record keeping. Again, web technology using an onboard aircraft server hosting an ASP site to parse and deliver content is a likely candidate.

Application of an Electronic Flight Bag

Electronic flight bag applications span simple, electronic 'page turners' to fully integrated systems that communicate with and can control other aircraft systems. Current hardware and software technology exists to cover the entire spectrum of requirements; all that lacks is a unifying architecture, set of implementation processes and a suitable regulatory environment. In an effort to deploy systems quickly however, some operators have defaulted to the lower end of the spectrum, opting for simple conversion of existing paper documents to electronic media using applications such as Adobe Acrobat enhanced with linking and bookmarks, as an interim measure. As these systems develop, however, operators will migrate from electronic products relying on paper-based structure and content, to integrated information systems that supply the information customized for the user's environment at the point of need, and simultaneously allow the information specialist visibility and control of the entire document content. This will lead to utilization of the self-assembling information concept.

An EFB device using the self-assembling information concept and processes will require tightly integrated web-based applications tailored to aviation requirements. For example, as the pilot reaches the notional 'top-of-descent' at Roanoke, he or she makes an appropriate input denoting the aircraft position, and the system returns prioritized, layered information consisting of the 'Preliminary Landing Checklist', 'Approach Briefing', and applicable arrival, terminal and contingency charts and navigation data, along with special information specific to that airport.

The pilot's input initiates a series of database, conversion and display actions that results in the information delivery. The EFB device translates this initial input into a structured database query that is further tailored by earlier inputs such as aircraft tail number. This query is passed to the database, which extracts applicable data, structures and sorts the extracted information, and then converts it to a document instance for display on the EFB. In addition, the information instance contains active pages or objects, allowing user-customizable options to further refine the information display for a dynamic flight environment.

The electronic documents returned by this system are not constrained by the same physical layout or structure as paper documents. This removes many of the paper-based limitations. This principal design paradigm is both a strength and challenge of electronic documents. The traditional concept of 'pages' dissolves as the user navigates logically through electronic documents via electronic links where pages may be created for the current requirement, but are not stored. Well-designed electronic documents further break information into complete objects by logical division rather than conventional structured paper methods, and do not require the user to simultaneously access several documents to gather sufficient information for a task. This is possible as electronic media storage capacity exponentially exceeds binder capacity for paper documents. Consequently, the author or program designs an electronic 'page' to optimize the targeted display characteristics, as opposed to a paper book's physical size. This is an important consideration not only for the cockpit, but also allows for the same information to be repurposed to other uses such as dispatchers or decision makers. However, this capability also results in challenges, as people are comfortable with the 'operation' of book-based documents, and regulatory approval processes must allow validation/approval of information based on a process other than numbered, dated pages. Progress has been made to date as there are draft Advisory Circulars in work to allow operators to use electronic record keeping without paper backup for critical data tracking. As the FAA continues implementing digital information approval regulations, page-numbering importance will decrease. The use of metadata within structured electronic documents is a possible option for identifying version and content to satisfy regulatory and operational requirements.

As vestiges of paper-based structuring disappear, a systematic navigation process will evolve; intelligent linking and navigation are critical. Direct links between sequential objects, cross-reference links to related objects, links generated by user-defined queries and menu selection links must be combined in a manner where navigation is intuitive and easily reversible. The EFB user interface must allow seamless transition between applications, and enable the user to shift easily and quickly back-and-forth from application to document. Fortunately, web technology has advanced to a point where this can be accomplished.

Summary

The structure of current size-limited paper-based documents requires the pilot to seek related information across multiple locations. In spite of electronic media's ability to rapidly navigate through documents, dispersed information still requires the pilot to keep information objects in mind while organizing them for a task. Air carriers must make a significant paradigm shift and utilize database and information display technology to transition to self-assembling information. Current cockpit documents must be decomposed into information objects based on a detailed task analysis. These objects, stored in a database, will be automatically assembled at a point in time at a specific location, to provide the user complete information for a flight or learning environment. These self-assembling information structures will provide prioritized, layered and linked information needed for current and subsequent related events, reducing task loading and leading to significant improvement in safety, efficiency and compliance.

References

Degani, A. and Wiener, E.L. (1994), 'Philosophy, Policies, Procedures, and Practice: The Four "P"s of Flight-deck Operations', In N. Johnston, N. McDonald, and R. Fuller (eds.), *Aviation Psychology in Practice* (pp. 44-67) Aldershot, UK: Avebury Technical.

NASA/FAA (2000), *Developing Operating Documents Manual*, Moffett Field, CA: NASA Ames Research Center.

Travers, R.W. (2000), 'Pilot Centered Phase of Flight Standardization', *Proceedings of International Conference on Human Computer Interaction in Aeronautics*, pp. 51-56, Toulouse, France: Cepadues Editions.

Chapter 8

Establishing a Shared Information Management System

Jack W. Eastman

Introduction

Establishing a shared document and information management system that meets the needs of numerous organizational and regulatory stakeholders is a daunting project. The extent of the conversion process required depends on how well the system is meeting current and planned operational needs. As with any complex system or process, the first of many steps involves determining what those needs are and how well the current system meets those requirements.

Over the years, the document management process at Atlas Air grew from that which supported a one aircraft operation to supporting a fleet of 37 B747 freighters. It would not have been feasible to create a document management system for one aircraft. However, eight years and 36 aircraft later, the original system left many information users frustrated.

In assessing the needs at Atlas Air it was (and continues to be) imperative that the operational and regulatory needs be thoroughly understood, and not constrained by the current and perceived process of distributing documents. 'The map is not the territory'!

The document conversion process at Atlas Air is changing the way information is structured, managed, standardized, shared and delivered globally. A chronology of events from the start-up of Atlas Air and its publication and document system, through the development and implementation of the vision, is offered as a context for the document management redesign process.

In the Beginning

For Atlas Air, the document conversion process started eight years ago when the company operation started with one aircraft. As a start-up air carrier, the easiest way to create or obtain the necessary manuals and documents was to have them created or converted by another carrier into a usable form for Atlas. For flight operations, the first flight handbook (the FAA required aircraft operating manual) for the first aircraft at Atlas was a derivative of another airline's flight handbook (FHB). Other operational departments have their own required documents and similarly adapted to meet their changing needs.

When Atlas acquired its second airplane, which was a variant model from the first, a 'differences' manual was created to highlight for the flight crews any different characteristics or equipment. The process of creating 'differences' manuals was born. Atlas continued adding aircraft and, by 1999, Atlas had one large FHB that represented the first Atlas airplane, and seven differences manuals covering the rest of the fleet. The eclectic manuals contained significant redundancies, were created in various software applications by a number of individuals, and distributed to flight crews after they were produced by being photo copied at a local (and most likely a rather profitable) copy center.

Data conversion covers a broad spectrum of activities and can be defined or classified at various levels (see Figure 8.1).

The most basic conversion in the airline publications context is the conversion of hard copy documents to electronic form by manually keying or scanning data into a word processing software application which may be either read only or read-write capable. At a higher level, one may convert electronic data from an obsolete operating system or word processing software application to a state of the art content management system. Another level is to convert a variety of publications produced by a variety of software applications into a standard form across the organization. In addition, conversions can be made to formats that permit web deployment of individual documents, web-deployed interactive documents, or web-deployed documents that are linked or interconnected such that multiple manuals can be accessed simultaneously through hyper linking. Finally, documents can be converted into basic content data such that it is no longer required to maintain separate files for each manual; only the data needed to create any or each organizational document and the protocol for organizing the content into a manual.

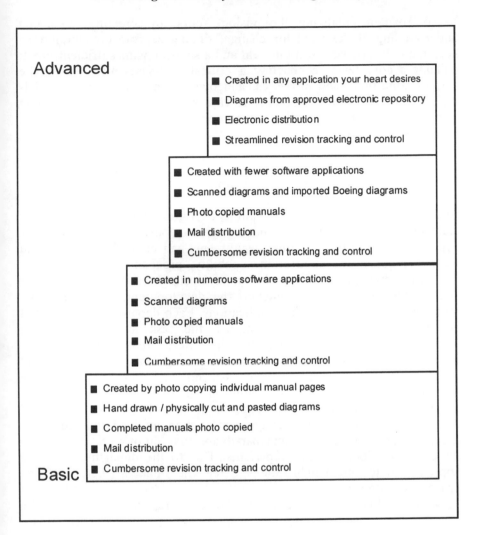

Figure 8.1   Levels of Data Sophistication

In establishing a common information or document management system, an assessment of current publication realities is necessary. Creating the right team to conduct the assessment is essential and leaving out key stakeholders can be extremely costly in terms of time and money. While all departments may share a common understanding of the problem and commit to co-creating an enterprise-wide solution, competing departmental goals inevitably create difficulties that are best solved by having a project sponsor with enough clout to ensure that these obstacles do not derail the project.

To road to realizing the vision starts by creating a shared understanding of the need for change; finding a project champion to keep the ball moving down the field and a sponsor with sufficient clout; building a core team; conducting a critical self-assessment as to the current conditions and processes; and overcoming the mental models we share regarding the problems it causes, solutions and establishing priorities. Only then can the actual conversion and implementation processes be undertaken.

Creating the Vision and Core Team

As Atlas Air's business expanded and its fleet grew, the requirements for manuals and documents expanded simultaneously. From the very beginning each of the operating departments had created and developed its own independent publication capability, procedures and processes. Each capitalized, as best it could, on its existing resources, experience and knowledge base as it related to the development, production and revision of manuals and other publications. Each department did what it needed to do in order to accomplish the task at hand. Early in organizational history, publications that were created electronically were produced in a variety of word processing software, in different versions, and on different operating systems, reflecting the knowledge and skills of the individuals creating them. As a result, within the organization a basic, eclectic and, as it turned out, dysfunctional document production and management system evolved that reflected the publication experience and paradigms that employees brought to the company from previous experience. Due to the department-centric approach to technical publications, a realistic financial picture of the processes did not exist.

Within this environment, it became apparent that as the organization moved forward and expanded its operations, there was a widening gap between the airline's rate of growth and the capabilities and capacity of its publications support. In some respects, we found ourselves 'hitting the wall'; that is, from 1993 to 2000, there was a widening gap between the publications system capability and the needs of the airline due to fleet growth. The ability to keep up with publications was deteriorating. Our publications history indicated that we would fall farther and farther behind if we did not adopt a new approach (see Figure 8.2). It was at this juncture that we realized a visionary approach and a program champion were needed as soon as possible. The dilemma we faced was that we needed to completely rebuild our publications processes while at the same time keeping the current processes afloat.

In that climate we realized that within each of our operating departments we needed publications process improvements in quantum proportions and set out to develop the vision. Initially, each department set out independently to attain publication process improvements within their department. From time to time efforts were made to coordinate manual revision processes between departments for those manuals that required cross-functional coordination; for example, the Dispatch Deviation Procedures Guide (DDPG) or the Aircraft De-Icing Manual, each of which requires close procedural coordination between flight operations, aircraft maintenance and ground services. As each department attempted to support another with electronic data, it became clear that if we had common or shared software, common or shared revision processes, et cetera, we could be significantly more effective within our individual departments as well as in overall organizational effectiveness.

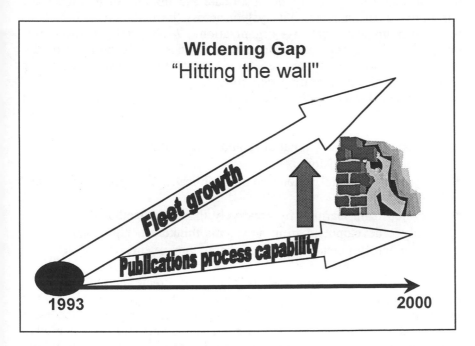

Figure 8.2   Increasing Gap Between Organizational Needs and Capabilities

At this juncture, someone realized the sheer magnitude of the potential positive, effective improvement that could be attained if we combined and coordinated our efforts. With that epiphany, a project champion was born who set out to communicate the impact this effort could have on the entire organization. Educating top management as to the overall value of a corporate wide, standardized and shared document management system was the first milestone. The project champion had the determination to stay the course throughout the project's conceptualization, planning, implementation and competing priorities.

Building the Core Team

Having gained top management support, the project champion formed a core team. It was clear that an organization's publications processes go well beyond individual operational departments like flight operations or maintenance. Therefore, the core team consisted of key members from each operating department, and members from the information technology (IT) and human resources (HR) departments. Input was sought from the purchasing and legal departments as necessary. The idea was that we were going to consolidate a common, standardized and shared document management system across the company. This would be a quantum step for any organization. What evolved from this perspective is a group called Document Services. This group provides document management services to all departments, while remaining independent of any operating department.

The basic charter of the Core Team is to accomplish and monitor the following:

- Focus energy, develop patience and view reality objectively.
- Create a shared vision.
- Overcome mental models of publication's departments and processes.
- Develop team learning by creatively thinking together.
- Analyze development from a systems thinking perspective.
- Continuously pursue of state of the art improvements.

Building the core team and its ability to create and develop the organization's vision of a system capable of integrating all of the requirements of our operating departments required two necessary steps. First, the team needed to loose the paradigms and mental models that provided limitations on how we learned and solved problems (organizational learning disabilities) and second, to conduct an organization self assessment of the current conditions and processes.

Overcoming Organizational Learning Disabilities

As we embarked on creating the vision, we came to realize that our current mental models had trapped us into living in a paradigm and not realizing how that thinking limited our ability to understand and solve the problems. We had every type of organizational learning disability. For instance, the enemy was always out there; there was always someone else to blame. We had a fixation to jump on an event when things around us were going wrong. When we were unable to grasp invisible gradual change, we would divide problems into segments, thus driving ourselves back into our departments. We convinced ourselves of the illusion of taking charge and the delusion of learning from experience when, in fact; we were creating or reinforcing our 'silos' and 'fiefdoms'. In short, we had a basic learning disability that could be summed up in 'I am my position', where responsibilities and actions were limited to the boundaries of one's respective position. But at some point during this process we realized the need to develop a shared vision. Once we had developed that idea, we could build a plan to create our shared vision.

Conducting a Self Assessment

The next step in the creation and development of our shared vision was to conduct a self assessment of our current document systems. For example, we found that there were considerable redundancies and incompatibilities resulting from having technical publication operations in three or four different departments within the company. In addition, we learned that it cost us $30,000 a month for different departments to ship materials back and forth to various stations around the world to maintain the individual aircraft on-board libraries for a fleet of 37 airplanes. Table 8.1 indicates some of the findings from our self assessment of the publication system within the organization.

One of the challenges would be to eliminate expensive and inefficient duplicate efforts across departments to the extent possible. We needed to create strong points of leverage by developing new ways of thinking.

Table 8.1   Findings from Self-Assessment of Publication System

Organizational Limitations	• Multiple departments independently producing and revising manuals. • Multiple departments following different approval processes. • No basic style guides for the company or the departments. • Limited inter-departmental coordination of changes. • Redundant distribution systems for aircraft publication supplies. • Limited general accountability or cost controls.
Processes Limitations	• No two departments using the same word processing software. • No two people in the same department using the same software. • Limited tracking systems as to revision status. • Publications were produced by photo copying methods. • Cut and paste common - often no electronic files at all.

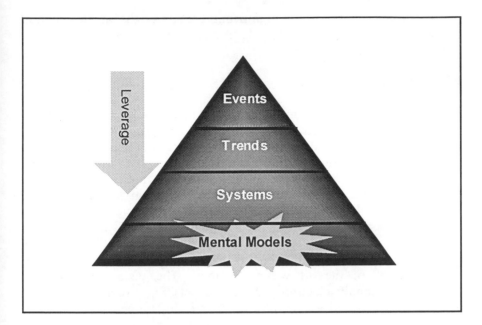

Figure 8.3   Elevations of Thinking

New Ways of Thinking

The most significant breakthroughs come when we challenge our thinking. The solutions we create are directly related to the level at which we understand the problem. As in Figure 8.3, the deeper we understand the problems, the higher leverage solution we will develop. Some of the levels of thinking are:

* **Events**- at this level we are attending to what has happened and are reacting to the event.
* **Trends**- at this level we are stringing together the events, connecting the dots and trying to predict the future based upon the pattern. This is helpful to the extent that the future is the same as the past.
* **Systems**- at this level we are looking even deeper as we identify the structure that creates the patterns we detected at the trend level. Systems and structure can be policies, physical constraints, programmed interactions.
* **Mental models**- at this level we are examining the thinking (individual and collective) that keeps the systems in place, which in turn creates the trends we notice and results in the events we endeavor to control.

We experience the most significant breakthroughs at the mental model level and therefore create the highest leverage solutions.

Elements of the Vision

Breaking away from a mental model that allows organizational fiefdoms to create isolated publications improvements, the vision of all electronic/digital manuals is applied across organizational boundaries; across flight operations, maintenance, ground operations, human resources, legal, et cetera. Our vision became the implementation of a global, state of the art document management system for managing all company manuals and procedures within an electronic environment. This would ensure that the right information would be available for the right people, at the right time and in the right format. Our objective would be to streamline and automate our documentation processes also by using state of the art web technology. The vision includes the capability to supply electronic document sets for aircraft operations and servicing. Our vision is to provide all flight operating manuals, performance computations and manuals, as well as navigation and approach chart data, electronically in the cockpit to the flight crews. This vision encompasses immediate electronic access to necessary documentation and data for maintenance and ground services personnel. In addition, other vision elements include:

- Full electronic document sets for aircraft operations and servicing.
- Direct link capability to all flight crews via electronic means.
- Direct link capability to all aircraft for data up and downloads.
- Complete integration and interactivity between and among documents.
- The ability to simultaneously revise and update complementary documents.

In general, the shared vision advocates continuously maintaining state of the art, standardized document processes and systems for now and the future.

Building the Management Plan

We think we have embarked on the right road in setting an enterprise-wide vision and in establishing the multi-departmental Core Team.

There are several critical tasks that must be accomplished in pursuing the enterprise-wide solution, including building a publications' matrix, conducting a content analysis and developing enterprise-wide layouts.

Build a matrix that lists all the organization's publications on the vertical axis and as many associated attributes as appropriate along the horizontal axis. Attributes should include, but not be limited to:

- Regulatory basis.
- Content authority (subject matter expert).
- Content owner (person directly responsible for the accuracy of the content).
- Associated software applications.
- Extent and type of graphics.
- Number of pages.
- Distribution requirements (method, recipient, etc.).
- Extent of interface with or impact on other documents.
- Source of revision requirements and formats.

Building this matrix provides invaluable insight as to the complexity of the task at hand and will develop a central point of reference for the assessment and coordination of document interface and conversion requirements.

There is a priority, or controlling system, associated with some operating manuals. For example, a change to the aircraft manufacturer's airplane flight manual (AFM) will drive changes to several operating manuals that will in turn drive changes to a set of dependent manuals and training materials. An analysis of the manual priority system and interconnectedness provides insights into the depth to which content in the manuals should be linked.

It is essential that the core team establish a set of priorities for the conversion process with respect to urgency and frequency of use of the manuals. The conversion of some manuals will be relatively simple and straightforward, while others will be more difficult. From this analysis, establish a conversion plan based upon set priorities. The best initial choices for conversions are those that will provide the greatest impact for the energy expended. If possible, 'hit a home run' with the first conversion, garnering the support that is critical whenever a change is introduced.

Conduct content analyses to develop document type definitions (DTDs) and style guides. Develop enterprise-wide layouts that specify the required formats and style guides.

Implementing the Plan

The plan, as outlined in Table 8.2, was developed in three phases. The first phase concentrated on general conversion and consolidation of documents. In the second phase, structure and management were added to the document system. The third phase addressed the delivery of documents via the Web.

Table 8.2 Implementation Phases

Phase One	• Systematically began the conversion process from hardcopy and several versions of Word and WordPerfect plus an assortment of graphics and spreadsheets.
	• Approximately 115 manuals initially consolidated into the Document Services Department in various media and formats:
	  • Microfilm
	  • Microfiche
	  • Tapes
	  • Electronic
	  • Paper / hardcopy.

Phase Two	• Converted digital manuals to PDF for Web deployment. • Acquisition and implementation of a document management system. • Development of full suite of DTDs as appropriate in conformance with ATA standard specifications. • Support development of ATA specifications for operation's manuals. • Systematic conversion of manuals to FrameMaker® with XML tagging. • Creation of a comprehensive content database to support the inter-linking of manuals to replace individual independent manual.

Phase Three	• Deploy manuals (as appropriate) around the world via the Web to:
	  • Aircrews
	  • Aircraft
	  • Maintenance crews
	  • Ground crews
	  • Customers
	  • FAA.
	• Interactive hyper-linking via the Web between and among all appropriate manuals for:
	  • Training center
	  • Dispatch and operations control.

Where are we Now and What have we Learned?

For the first time in Atlas' history, we have digitally published a flight handbook. Our previous standard practice was to copy a flight handbook a thousand pages long that cost us about $75 a copy plus shipping. When we finally printed a handbook recently, it cost us $24, shipping included. That is a huge saving. But continuing the vision, we want to have all digital electronic manuals, not just for flight operations, but for all departments in the company. Also, we are on board with the ATA Flight Operations Working Group in support of industry-wide standards. In the conversion process of our flight handbooks, we have adopted the ATA standard specifications for aircraft systems numbering. Further, we are following the evolving ATA standard practice of phase of flight specification to the extent we can. If all airlines followed a standard such as the ATA specifications, then the differences between their operating manuals could be dramatically reduced. This level of standardization would benefit the industry in a number of areas such as:

- Ground and flight safety.
- Ease of learning from one company's manual to the next (i.e., job change).
- Performing contracted services to another airline.
- Communications from aircraft manufacturers.
- Communications with domestic and international regulatory agencies.
- Simplifies finding information in manuals with different languages.

Technology has enabled potential quantum process improvements and there will be continual change. In this context we must continually evaluate technological improvements and be keenly aware of our mental models and their impact on our thinking. In addition, we need to be smart and realize that one size does not fit all. At the risk of sounding cliché, we have learned that this truly is a journey, not a destination. Technology is changing by the minute and, with strong industry-wide coordination and partnering, we can collectively challenge our thinking and create systems that previously would have seemed impossible.

Reference

Senge, P. (1990), *The Fifth Discipline*, NY: Doubleday.

Part 3
User Innovations in Aviation Operational Information

Chapter 9

Electronic Flight Bag in Action: The JetBlue Experience

Brian L. Coulter

Operational Philosophy of the Electronic Flight Bag

During the certification of JetBlue we realized that we had a unique, albeit dichotomous, publications opportunity available to us. As a new company we had no legacy system to overcome, but conversely, we had only a few months to develop a viable manual system. The certification team had a lot of previous experience with paper systems, but little experience with the employment of an electronic solution. One of our most important choices was the selection of an appropriate media for transmission. We knew we wanted to make use of the computer network that we were building, but initially there were no electronic solutions available on which to base a plan of action. This lack of a working model required us to develop our own philosophy.

The certification team's experience with paper-based manual systems provided a good basis for being able to present a worthwhile business case to upper management in support of an electronic solution. This business case was somewhat more difficult to make when dealing with a legacy paper system, but there was significant evidence to suggest that an electronic solution is still more cost-effective.

Although we quickly became convinced that an electronic solution was viable, we still needed the agreement of our FAA certification counterparts in order for the system to move ahead. As we were the first carrier to ever be fully certificated under the Certification, Surveillance, and Evaluation Team (CSET) concept, we also required concurrence at a national level as required by HBAT 99-09 and HBAW 99-11, Introduction of an Automated Process for Certification of New Entrant 14 CFR Part 121 Air Carriers (FAA, 1999). Our Principal Operations Inspector (POI) and our Principal Maintenance Inspector provided considerable insight into the FAA requirements. Additionally, the CSET leader and members provided additional guidance from the national perspective. This information was invaluable, allowing us to design a system that would be compatible with existing FAA information.

Our original plan was to design a single manual, as referenced in the FARs, and linking this single source to all of our users. We quickly saw that this approach was not feasible. The biggest obstacle to developing a single manual was the need to have the airline certified in six months. While this approach may have been viable for some of our internal manuals, manuals that required a significant input from the manufacturer could not be easily introduced or updated. This lack of consistency in database information between internal and supplier manuals is one of the major hurdles to a true electronic manual system. It was not a total obstacle, but it did constrain the overall project. We did not have the resources to design a sophisticated system to archive the required information with the functionality we wanted. Thus, the basic requirements of the paper system were to remain, and we would not be allowed to change document:

- Format.
- Content.
- Organization.

This went a long way to determining the page layout of our manuals. They would have headers and footers, page and revision numbers and a visual system for indicating revisions. The system would also have to be as reliable as current paper systems. In fact, stringent tests were made to ensure that the principles of Distribution, Content, Revision and Access were all at least as good as a paper-based solution.

We also had to cater to mobile users of the manual system. The principal targets of this group were the pilots. They had the greatest requirement for up-to-date information and would be spending the most time away from our main network. Other mobile users included Inflight and Quality Assurance crewmembers. Each of these groups would require a slightly different solution (see Table 9.1).

In addition to the official requirements, we also faced a substantial technical challenge. Based on the planned concept of the computer network, we were required to develop a distribution and revision system in the shortest possible time.

Table 9.1   User Requirements and Solutions

Group	Mobility	Reference Access Required	Power Source	Solution
Pilots	Mobile	Constant	Aircraft	Laptop
Quality Assurance	Mobile	Constant	Terminal	Laptop
Flight Attendants	Mobile	Periodic	Not Available	Paper
Maintenance Technicians	Static	Periodic	Terminal	Workstation
Customer Service	Static	Occasional	Terminal	Workstation
Administrative	Static	Occasional	Office	Workstation

The Original Plan

Direction

In order to map our direction we tapped several sources. We collected and read as much information as the FAA could provide on the subject of manuals. This was augmented internally by the knowledge of two key individuals, who provided a combined 70 years of aviation experience in both flight operations and maintenance. Moreover, they had been involved in the certification or re-certification of some 14 different airlines. We then started with the regulations and backed those up with the interpretations provided by Advisory Circulars (ACs) such as AC 120-69, Use of CD ROM Systems (FAA, 1997), and Chapter 15 of the FAA Order 8400.10, Aviation Transportation Operations Inspectors Handbook (FAA, 1994). It became evident that the FAA had begun dealing with the implications of an electronic document system. Certainly, maintaining a manual in an electronic format was supported. However, the guidance available when we started JetBlue was largely dedicated to CD-ROM-based solutions. While these were helpful, to some extent, they also required some formatting and distribution limitations left over from paper-based documents, which later proved to be significant hurdles to electronic implementation.

Off-the-Shelf System

We decided that the use of off-the-shelf systems was paramount. With the help of our entire IT department (consisting of a single individual), we set about finding systems that would support our goal. After a long iterative process, we found systems that we could connect together easily and satisfy all our needs. In fact, our initial system was fully functional on a total of four lines of code. Once we had proved to ourselves that a system functioned correctly and was reliable, we demonstrated it on a number of occasions to both CSET and the certificating local FAA office.

Avoiding the Traps

One of the anomalies of paper-based systems we were trying to avoid was tied to a distribution of revisions. Traditionally only an owner of a manual would actually physically receive the revision. While this suits most groups, it certainly does not suit some of the ground-based groups where sometimes only the office staff actually possesses a copy of the manual. In our system, everyone who uses, or is required to have knowledge of, the contents of a particular manual is sent a Revision Notice. This Revision Notice contains a response button that requires the user to acknowledge the receipt of the revision information. While the new manual was not included in this e-mail, a detailed description of the change in procedures was included. This meant that no matter what the grouping, all users received information on any manual change. One of the early technical challenges of this system was that only Microsoft Exchange supported this acknowledging system. This, in turn, could only successfully operate on a Windows NT based system. Thus, the requirements of the Revision Notices drove the requirement for the operating system.

Another paper-based anomaly we hoped to bypass was the lack of fidelity across the manual system. Although a database style single manual where all users were accessing a common source could have gone a long way to achieving this aim, as discussed earlier, we did not feel this approach was feasible under the circumstances. We therefore went to considerable lengths to design other processes that could help assure this fidelity. We started by creating a manual on manuals (MOM) that was colloquially known as the 'mother of all manuals'. Within this manual we included a section dedicated to manual interface. Included in this section was a very important table known as the interface matrix. This interface matrix provides a reference for all known manual interfaces down to the subject level. All persons responsible for revising manuals could use the matrix to establish any overlaps in the manual system. Moreover, the matrix provides a sound basis for the design of any future database style manual systems.

System Design

Simplicity and Security

Simplicity and security were the underlying concepts in designing the system and they were largely solved by the choice of a display format. The two leading contenders were HTML or Adobe Acrobat documents. They were both already widely used on Web applications and would require little extra training. However, Adobe Acrobat provided a much greater level of security. Adobe had a relatively easy method of saving a file that effectively rendered the document unchangeable to all but the most advanced hackers. Additionally, the system could easily use its built-in cryptographic keys or plug in almost any other security system. Lastly, Adobe Acrobat provided a page layout that was identical to a paper-based system. This meant that we could eliminate almost all of the layout questions that the FAA had very quickly. It also meant that if at any time our electronic transmission request was denied, we would have been able to print our manuals directly and start a paper-based system. At the time we put our system together, it seemed too cumbersome to introduce a fully functional SGML system. This would probably have produced the best results, but would have required a lot of extra IT resources to accomplish. HTML was certainly a more cost-effective solution, but was already being touted as out-of-date because of its inherent limitations. XML was only starting to be used, and was an unknown, albeit promising, format. Thus, with the XML family in doubt, we went with the best option available.

Simplicity and security also had to extend to the updating system. We could not require the user to have an in-depth knowledge of computer systems to start or finish an update. The most logical process to tie updating to was the user logon. This would already require a user to have very basic computer skills but would not tax those skills any further. The only additional skills required by our users are the ability to send and receive e-mail, and the ability to read information from a computer screen.

Transparent Updating

Although the user updates were tied to logon, the system updates were accomplished based on time. We used the natural down time of the early hours of the morning to accomplish a system wide update. In this way, we were able to monitor the distribution of updated material and ensure consistency across our network. If an update failed, an automatic e-mail message was sent to the IT department. Additionally, each time an update occurred, the system would write a message of its success and destination. This provided both a positive and negative means of recording the updates.

Operational Issues

The Quick Turn

As JetBlue was intending on performing 30-minute turns at all stations, we needed a manual system that could support that requirement. The first plus of the electronic system was that the pilots effectively carried a complete set of company manuals at all times. That meant that we had additional resources available in the cockpit area to help solve maintenance issues and outstations. These resources proved invaluable in both the certification proving runs and during the early phases of JetBlue's operations (see Figure 9.1 for a picture of the EFB in the cockpit).

Figure 9.1   EFB in Use in JetBlue Cockpit

Operational Flexibility

In the dynamic world of commercial aviation, sometimes seemingly minor changes to procedures can have profound effects, the most important of which affect safety. The ability to reduce the distribution of these procedural changes from weeks to days and sometimes even hours, had very positive effects on our operational flexibility. In a legacy system, the crewmember is sent a copy of the revised pages either by mail for a dispersed operation or by the crewroom/v-file system for a hub-based system. The crewmember is also still responsible for actually including the new information in the appropriate book. The crewmember is provided a return mailing or drop box page for later collection. These responses are collected by technical publications and are collated to ensure crewmembers are adequately informed. This can take up to three weeks, even if the deliveries go well. The electronic system will automatically update the information used, and has the ability to distribute the information and retrieve the responses in hours rather than days or weeks.

Reliability Issues

Laptop Reliability

One of our principal concerns with the choice of hardware was how the operating environment would affect the reliability of our laptops. Our initial choice proved to be suitable for all groups within the company except pilots. At one stage, some 10 percent of our pilots' laptops were undergoing some form of maintenance. Mostly these were hardware issues with either the keyboard or screen. However, some software and operating system incompatibilities also caused problems. We chose to replace the original hardware with a different model that was both a little more expensive and reliable. Ironically, the original laptops are still running extremely well for other mobile groups within the company. We have not conducted any significant tests to establish the reasons for this anomaly.

Power Supplies

Before deciding what power source we would use, we canvassed European operators that were using the laptops in the cockpit for other applications. Based on their recommendations, we decided to use the solid-state transformer provided with the laptop rather than modifying the aircraft's power supply. This not only saved time in certification and completion of the modification, but also provided operational flexibility by allowing the change of laptop without a change in aircraft status.

Network Maintenance

Having an IT department involved in the process has also proved vitally important. As our network evolves there are many incremental decisions that could ultimately affect the ability of the network dependent functions to operate properly. Only by maintaining a complete overview of these network dependent functions has the IT department been able to avoid major pitfalls. Additionally, by maintaining an overview of network traffic, certain portions of the updating system have been modified to make better use of network capabilities.

Usability Issues

Transfer Tasks

Since the inception of our electronic cockpit, we have been working to reduce transfer tasks. These quite literally are tasks that require crewmembers to transfer the information from one electronic source to another. These repetitive transfer tasks are prone to error. We were not aware of the magnitude of either the tasks or the errors until this issue was reported to us. Transfer tasks were highlighted as a part of an Airbus Industrie study into the use of laptops in the cockpit. The results were not published, but the information was used as part of the Less Paper Cockpit presentation at the Airbus 11th Performance and Operators Conference.

From observations and recommendations from the study, we have made it a point to try to reduce transfer tasks to an absolute minimum. Most of these tasks are necessary because our aircraft are not as yet part of the JetBlue network. The ability to connect our aircraft to the remainder of the network has taken a much higher priority since the study was conducted.

Quality of Words

Another of our most significant findings in the area of operational usability is that, no matter what the chosen media, the quality of the words included in the manual system is of the utmost priority. The ability to guarantee the fidelity of a procedure in a manual, and the ability to update it at a moment's notice, means little if crewmembers cannot interpret its message. The quality of the received message must be constantly monitored so that crewmembers can understand what is required and accomplish the appropriate actions.

Acceptance and Recruiting

Before any medium can be considered successful, it must be accepted by the crewmembers that use that medium. One advantage JetBlue has is that the use of computers and their associated systems is required throughout training. This requirement means that new crewmembers use exactly the same media during training that they use operationally. It has been anecdotally observed that this cuts down significantly on transition problems from the classroom to the job.

At this stage we have found no magic formula for a crewmember's adaptability to the electronic arena. Previous computer experience and exposure are definitely helpful, but not a complete panacea for electronic adaptability. We have made several modifications to our selection criteria to assist in forecasting this compatibility. While we think we have made positive steps, we have still not found the whole answer.

Lessons Learned

Functional Flexible Software

One of the keys to making an update system successful is the choice of the right software. Not only must the software be functional and achieve its primary goal, but its design must be modular enough to enable timely modification. In one example, we programmed our software to display certain messages received from the aircraft. This worked well until we realized that there were other messages we needed to display in order to reduce transfer tasks. Due to a bottleneck at the software supplier's end, it took six months for the modifications to be made. At all times the software was functional and achieved its primary goal, but the time required to modify the system was unacceptable. This may lead us to change software earlier than anticipated. This really highlights the need to not only purchase software that fits current needs, but that also has the ability to adapt to changing needs. This is best done as part of the competitive negotiations prior to purchase.

Gate Connections

As alluded to in earlier sections, JetBlue is working on the ability to connect our aircraft to our network. Initially this will occur only when the aircraft is on the ground and within range of a gate transmitter. While these gate systems require a not-inconsiderable capital outlay, they will enable us to economically transfer the large packets of data at a much more cost-effective rate. This is one of the key enablers in our aim to reduce transfer tasks. Additionally, this will also provide a more secure means for transferring sensitive data between the ground and aircraft. Moreover, the goal of achieving FOQA analysis in real time also may be a step closer.

Airborne Information

A longer-range plan includes the ability to use some satellite bandwidth to provide pertinent pieces of operational information to the flight crewmembers when airborne. Such information might include:

- Updated weather analysis.
- Updated radar images.
- Updated wind values.
- Expanded recommendations in routing and cost savings.

The cost benefit analysis on these features is ongoing, but is showing considerable promise. A less obtrusive benefit of maintaining a continuous aircraft connection is the ability for the pilot group to have an entirely consistent interface on their laptops. This apparently seamless approach would remove the conditional nature of the availability of some forms of information. Once again this goes directly toward reducing the number of transfer tasks.

Summary

The Electronic Flight Bag (EFB) concept is provided for in the regulations and current supporting material, but operators are burdened with the task of convincing their inspectors that the approach has the currency, consistency, availability and distribution characteristics of the legacy system. However, a 'true' EFB will not be feasible without an industry effort to agree on database definitions. This will allow airlines to individually customize the user interface, but be able to universally use the information provided directly from the manufacturer.

References

FAA (1994), *Air Transportation Operations Inspector's Handbook, 8400.10*, Washington, DC: Federal Aviation Administration.

FAA (1997), *AC 120-69 - Use of CD-ROM Systems*, Washington, DC: Federal Aviation Administration.

FAA (1999), *HBAT 99-09 and HBAW 99-11 - Introduction of an Automated Process for Certification of New Entrant 14 CFR Part 121 Air Carriers*, Washington, DC: Federal Aviation Administration.

Chapter 10

Design and Certification of an Integrated Aircraft Network

Robert K. Bouchard

Introduction

One of the greatest challenges an airline faces is managing information. The key objective is getting information to individuals where and when they need it in a format that can be used by everyone. Federal Express (FedEx) has a project called 'TITAN' that meets this challenge in innovative ways. The Totally Integrated Technical Aircraft Network (TITAN) project is a combination of networking components and input/output devices that allow for automatic and interactive use of data and information. FedEx's strategic long-term vision is to eliminate most of the paper documents on the aircraft, to provide high-speed, low-cost, realtime communications between the aircraft and the FedEx ground network, and to create an electronic interface between the aircraft airworthiness log book and the FedEx maintenance mainframe system. To accomplish these goals, FedEx worked with its vendor (Spirent) to develop an infrastructure on the aircraft that would allow for growth in the years to come. It was realized that once these baseline components had been installed and certified, adding additional components and functionality could be done at less cost with few certification hurdles.

The user interface was designed using 'the weakest link' concept. Program the software such that the system is as intuitive as possible so that minimal training is required and the most computer illiterate person can use it. The motto was; 'Any steps that can be engineered out so it is easier for the end user, is worth doing.'

The certification of the 'TITAN' system was managed in little pieces. By splitting the TITAN system into chunks and presenting them to the Federal Aviation Administration (FAA) in understandable pieces, certification of these components became possible. Installing file servers and laptops using commercial off the shelf (COTS) hardware and software on the aircraft was a totally new concept to both the operators and the FAA. It was important to work very closely with the FAA to make sure both parties understood that what was being developed was in adherence to all the FARs and specifications.

We have learned that, due to the speed of technology, building and certifying robust traditional avionics systems is cost prohibitive. The time it takes to design, build and implement a network system using 'traditional avionics' would cost too much and never allow for an affordable upgrade path as technology progresses. Long design times, long lead times on parts and high costs of design and manufacturing are avoided by using COTS equipment. There are drawbacks as well; for example, it is almost impossible to obtain components that meet every desired specification and sometimes compromises must be made. Still, the advantages of the COTS architecture far outweigh the disadvantages.

Integrated Aircraft Network

In 1993, FedEx was approached by McDonnell Douglas and Computing Devices International (currently Spirent Corporation) about a joint venture to automatically extract digital aircraft fault data from the on-board computers and electronically send that data to the ground. FedEx was intrigued and joined the team. FedEx would only join the team if we were able to help design the user interface specifications as the end users of the system. Thus a new system was born called the On-board Maintenance Terminal (OMT) system. This system would automatically pull the fault data out of the Central Fault Display System (CFDS) and down link them to the ground using ACARS.

The On-Board Maintenance System

The OMT provided the groundwork for what is currently called the 'TITAN' system. As mentioned, the design of the system was such that expanding its capabilities by adding external components would be easy and fairly inexpensive. The OMT Server had a firewall that was already linked to ACARS Controller and the Central Fault Display Interface Unit (CFDIU). The OMT also had an Ethernet Hub that connected to the OMT Display Unit in the cockpit. After more than three years of designing, bench testing and flight testing, we obtained certification in early 1997. Getting the aircraft computer fault data has been an enormous help to FedEx and has prevented many delays. It has allowed FedEx to accurately predict and diagnose many aircraft failures, saving money and time.

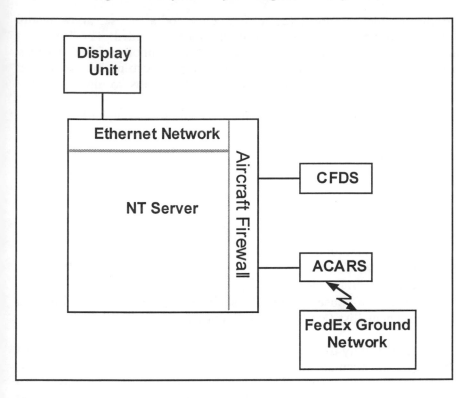

Figure 10.1 The On-board Maintenance Terminal System

Figure 10.1 illustrates the OMT system, which was the first phase of the TITAN system. Knowing what the computers were seeing was a great first step, but we were still missing crucial input from the flight crews. Just because a computer registered a fault does not mean a crewmember would notice a problem with the aircraft and visa versa. The next obvious step would be to develop a device that would allow a crewmember to accurately inform the ground personnel of in-flight problems so they could prepare for maintenance before the aircraft landed. The objective was that if you knew exactly what the pilot was seeing and exactly what the on-board computers were seeing (through OMT) then you could very accurately troubleshoot an aircraft failure.

The Pilot Access Terminal and Other Aircraft Components

The device to send crew messages to the ground is called the pilot access terminal (PAT). Again using the 'weakest link' theory, we designed the PAT as a point and click system using pictures to drill down to a very specific problem. A crewmember could drill down to a specific problem within three to four clicks of a mouse, allowing the pilot to add extra text when extra information is needed. The PAT ties into the already established Ethernet network of the OMT Display Unit to talk to the OMT server and uses the already established connection between the OMT Server and ACARS to downlink the messages.

The PAT being a COTS Laptop running a standard Windows operating system allowed for added functionality without adding cost of additional hardware on the aircraft. Several other functions have been integrated or are in work for the PAT:

- Performance calculations.
- Fault reporting.
- Pilot/mechanic electronic logbook.
- Flight/ground communication through server connection.
- Platform for other applications (e.g., Weight and Balance Program, Fueling program, viewer for Jeppesen Data).
- Host for in-house documents such as Flight Manual, Minimum Equipment List (MEL), Flight Operations Manual.

FedEx currently uses a HP Notebook monochrome laptop on all of the fleets to calculate performance. It is an Intel 486/25 processor computer that operates off of four AA batteries. Besides having to constantly replace batteries, another problem with this unit is that components for this device are no longer being manufactured. A new device is needed to run this application. The PAT, having a rechargeable battery and running a standard Windows operating system, allowed for this replacement.

The largest return on investment will be realized by the 'Fault Reporting' application on the PAT. Sending exact problem reports to the ground when they happen in flight will allow for advanced troubleshooting and maintenance planning. Having the ability to plan ground time, plan maintenance and schedule aircraft to minimize delays will change the way FedEx optimizes its resources.

The 'Fault Reporting' application ties into FedEx's desire to transition to a 'virtual paperless cockpit'. The application allows for the pilot or mechanic to enter log book discrepancies electronically and transport that data to the ground maintenance computers electronically. Because the log book write-ups are created electronically, the structure of the messages is consistent. Many departments within FedEx will benefit tremendously from having consistent data. Database search tools can create automated reporting for Engineering, Reliability and Maintenance Control departments when the data are precise.

With the PAT being connected to the server and the server having the ability to transmit data to and from the ground, we have increased the communication capabilities between the aircraft and the ground immensely. The user interface is such that the PAT keyboard makes it easy for a crewmember to type a message and direct it to a specific end user. Based on the priority, this can be forwarded to an individual's computer, e-mail or even a pager or cell phone.

Besides being a platform for the electronic logbook and electronic documents, it will be the platform for several other applications such as the FedEx Weight and Balance program, the Fueling program, and a viewer for the Jeppesen data. With the FedEx server architecture in place, the sky is the limit on the applications that could be hosted on the PAT, server or any other component of the 'TITAN' system.

Several of the documents that FedEx creates in house, like the Flight Manual, Minimum Equipment List (MEL) and Flight Operations Manual, can be hosted on the PAT. Having the capability to display these manuals on the aircraft removes the requirement for the pilot to actually carry these manuals with him or her. When you consider that the average flightcrew bag weighs between 25 and 40 pounds, any reduction in weight is welcome. The same applies to the Jeppesen charts that have to be carried by the pilots. FedEx is working with Jeppesen and Spirent to integrate the Jeppesen chart data into a user-friendly graphical user interface. Jeppesen would provide the raw data and Spirent would create a custom user interface. Having a server on the aircraft that is tied into the aircraft flight management system (FMS) would give us the capability to link the Jeppesen charts to FMS data. Imagine entering the departure/destination airports in the FMS and the PAT configuring the Jeppesen user interface based on this aircraft data.

Planning the certification of the TITAN system started with creating a firewall between the aircraft systems and the OMT server. Separating the server software from the aircraft systems with a firewall that only permitted specific commands to pass through the firewall was the objective. This made it easier to work with the FAA when certifying the equipment. Although it was quite a challenge to build the firewall and prove non-interference between the server (and the additional TITAN components) and the aircraft components (ACARS, CFDIU, FMS, etc.) it was worth the time and effort.

Figure 10.2 illustrates the addition of the PAT and a Full Format (8.5 by 11 inch) TCP/IP Graphical Printer. Figure 10.2 also illustrates the addition of other aircraft components that can be attached to the server via its Arinc 429 bus or Arinc 573 bus (Arinc 717 for digital flight data recorders). This allows us to listen to data from on-board computers such as the flight management computer (FMC), the traffic collision and avoidance system (TCAS), the global positioning system (GPS), etcetera, and integrate these into specific applications on the server and the PAT.

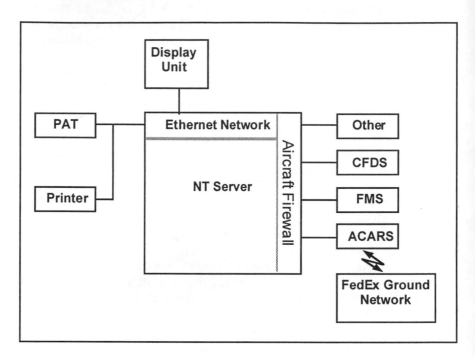

Figure 10.2 The On-board Maintenance Terminal with Pilot Access Terminal, Printer and Server via ARINC

Other Ethernet Devices

The TITAN server has two network cards and an Ethernet hub. One network card controls the aircraft network through the Ethernet hub (devices like the PAT, printer, OMT display unit, etc., are part of the aircraft Ethernet network). The second network card controls the Gatelink network and connectivity to the ground network. This is where the robust design of the server has paid off. Having the capability to add Ethernet network devices to an existing system has greatly increased the capabilities of the TITAN system.

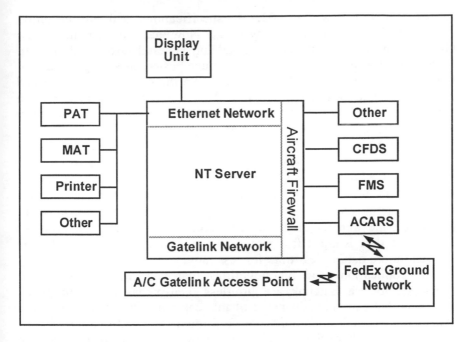

Figure 10.3 Total Aircraft Network

Figure 10.3 illustrates the addition of other Ethernet devices such as additional PATs, additional printers or a maintenance access terminal (MAT). With the Ethernet hub on the server, it is easy to add additional devices that can network with each other. Figure 10.3 also illustrates the addition of the Gatelink system. The Gatelink system will allow for high speed data transport between the aircraft and the ground network while the aircraft is parked at the gate.

Gatelink System

Gatelink is an industry term for high speed communication between the aircraft and the ground network using many means of transport while the aircraft is at an airport gate. Examples of these means range from plugging an actual Ethernet cable into the aircraft, to Infra Red, to wireless radio. Because FedEx was provisioning its Memphis hub with wireless radio access points (radios) for the tracking of cargo containers, managing its fueling distribution and for connection of a wireless laptop for the aircraft mechanics, FedEx opted to use wireless radios as its means for 'Gatelink'. We took advantage of an already existing system and added our aircraft to that network.

The radios follow the 802.11b specifications and are currently running at the 2.4 GHZ frequency range. Now that we are going to use Gatelink on the aircraft, FedEx is adding Gatelink Access Points to several other airport locations. This is much easier for a cargo operator like FedEx because, in most cases, they own their own ramp space and park in the same locations every time. This would be much harder for a passenger carrier because they normally do not know which gate they will be going to when they go to a new destination.

Another major benefit to Gatelink is that once you build the ground infrastructure, the data transfer is virtually free. When using the traditional methods of aircraft data transportation such as SATCOM, ACARS, and now even cellular, you wind up paying for that service either by the message or per minute. With Gatelink, once the network is in place, the transfer of the data is free. Not only that, the transfer rate of the data is much faster than the other methods described. You can send more data in less time for less money.

The applications and functionalities that can, will and are being developed to use Gatelink are limitless. FedEx has many proposed programs that can take advantage of this communications link that will change the way hub operations are performed. We feel that this is just the tip of the iceberg though. There are new suggestions being made on a daily basis.

Certification of the Gatelink system was fairly easy because we were able to show that Gatelink was only active outside of the flight realm. This means that Gatelink could not interfere with any of the aircraft systems because it was only turned on after the aircraft had finished its flight phase. The system is turned off again when a new flight phase is started. FedEx controls this through the Door Armed switches in the aircraft main entry doors. A flight ends when the doors are disarmed and a flight starts when the doors are armed triggering a relay that turns Gatelink on or off.

Designing Ground Network Functionality

Figure 10.3 shows how the aircraft side was developed and implemented. To obtain full TITAN system functionality, there is another major piece that has to be designed and implemented. The ground network is just as important as the aircraft infrastructure. The power of the TITAN system is not really on the aircraft side; rather it is on the ground side. It is the difference between just plain data and information. It is what you do with the data that makes the system most valuable. Figure 10.4 illustrates what is needed on the ground side that allows the aircraft equipment to function as intended.

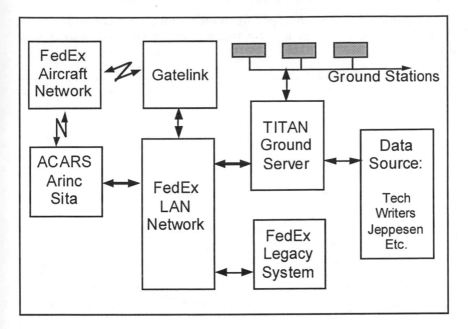

Figure 10.4 Ground Network

Moving the data to and from the aircraft is also illustrated in Figure 10.4. When the aircraft has data to send to the ground, or when there is data to be sent to the aircraft, the system first tries to establish communication with the ground through Gatelink. Only if this is not possible will the system try using ACARS as the alternate means of sending certain data. Due to the bandwidth of these systems, sending large amounts of data is only possible with Gatelink. However, the software of most of the applications in the TITAN system is developed such that most messages can be formatted in a single ACARS message (approximately 250 characters per message).

This is where we get a great cost saving. Currently, approximately 25 per cent of all of the ACARS related messages are sent to and from the aircraft while it is parked at the gate. Now that we are using Gatelink to send these messages, we have the opportunity to save up to 25 per cent on our current ACARS bill. This is of course dependent on how many stations are Gatelink capable. FedEx's intent is to eventually have every one of its airports outfitted with Gatelink radios.

Once a message has made it to the ground either through ACARS or Gatelink, it is passed on to the FedEx Intranet. Here the data is parsed according to message type and passed on to respective systems. In many cases, the data is sent to several locations. Figure 10.4 illustrates possible message traffic flows. In all cases, the data is available to the departments who need the specific information. As shown in Figure 10.5, the onboard system is but a fraction of the ground system with which it interacts.

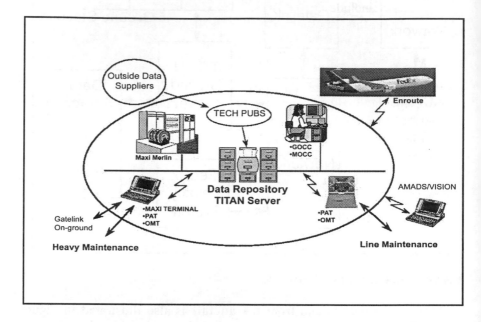

Figure 10.5 Major Components of the TITAN Project

Development and Certification

When developing the TITAN system, the following guidance was used throughout the process:

- Certify the system in bite size chunks, so the concept becomes manageable and understandable to management and regulators.
- Make sure that you include the FAA in every aspect of the design. Present data to the pilots, mechanics and other users in a way that is easy to understand.
- Use as much COTS equipment as possible.
- Re-use as much of the existing technology and architecture as possible.
- Manage all data through configuration control to maintain accuracy.

A key to our successful strategy was to develop and certify the system in small chunks or modules, so the concept becomes manageable and understandable, especially to the regulators. As part of this, we made sure to include the FAA in every aspect of the design. If we had not followed this approach, we could have spent a lot of money designing something that was not certifiable in the FAA's eyes. Since technology is evolving at a rapid pace, it is important that the FAA stay informed throughout the design process.

On the user interface side, displays were designed so they were easy to use for both pilots and mechanics. Regarding hardware, several strategies were used to help manage cost and reduce risk. A concerted effort was made to use as much COTS components as possible. COTS can be cheaper, easier to obtain and may have better warrantees than avionics systems. Through the use of COTS, we avoided long design times, long lead times on parts and high costs of design and manufacturing. Further, during TITAN's evolution, much of the technology and architecture were reused. For example, FedEx piggybacked on the Access Point Gatelink architecture being deployed for other departments. Finally, when working with operational information, accuracy is essential. We used rigorous configuration control to manage the entire document system to include the databases and manuals. These guidelines combined to allow FedEx to be one of the early adopters of an integrated aircraft network for operational information.

Certification of an Electronic Flight Bag

The draft Advisory Circular 120-EFB (FAA, 2001) outlines the process that FedEx followed with its PAT/TITAN design. The Advisory Circular has not been finalized, but after reviewing the formal draft, it looks promising as a tool for planning and certifying electronic document access and display systems in the cockpit. If released, it will give the Aircraft Certification offices direction on how to help operators certify these types of systems on their existing fleets, as well as guide aircraft manufacturers and their respective FAA offices for certification on new aircraft.

Learning from the FedEx Experience

As the development of the TITAN system continues, several of our guiding principles have proven to be productive strategies. First, start with a small target and build from there. Find out what other departments in your organization are doing so you can integrate using existing technologies and philosophies. Make the system as user friendly as possible by designing the system with the end user's participation. It is more likely to be accepted when designed by the user. Finally, work closely with your regional and local FAA from start to finish.

Reference

FAA (2001), *Draft Advisory Circular 120-EFB, Guidelines for the Certification, Airworthiness and Operational Approval of Electronic Flight Bag Computing Devices*. Washington, DC: Federal Aviation Administration.

Chapter 11

Display of Electronic Information in the Cockpit

Daniel R. Wade

Introduction

This chapter addresses the display of electronic information in the cockpit and the process used to assess various Electronic Flight Bag (EFB) and surveillance system needs and options in defining how best to display operational information. This is based on an industry effort, the Pilot Information Display (PID) Working Group, formed to plan and implement an EFB concept that enables optimum information management for pilots, maintenance and operations personnel.

In order to best meet the needs of the user groups and to stimulate industry-wide cooperation in establishing an efficient information management system, representatives from the different groups helped define operational information display and control requirements in the cockpit. The result was valuable input received from operators, operator based working groups, pilot focus groups, aircraft manufacturers and data/service providers.

The evolution of this EFB from an industry concept into a tangible, certifiable display is outlined below as a success story and road map for future advanced technology industry initiatives.

Pilot Information Display (PID) Working Group

In late 2000, Astronautics undertook to form an industry working group to define and develop requirements for an electronic information and display system that would meet the needs of operator EFB requirements. A primary goal was to optimize the display of electronic information in the cockpit. Realizing that the development of any new product must meet a market demand in addition to complying with perceived Advisory Circular requirements, Astronautics started with a blank sheet of paper and formed a working group consisting of various operators and suppliers. A series of working group meetings were conducted with the goal of defining what the target users actually required and what human factors considerations had to be met. Initial discussions addressed lessons learned from past and ongoing projects, future functional requirements, human factors concepts, installation considerations, training, data management, certification, cockpit display specifications and desired applications.

The first topic addressed was the lessons learned from past and ongoing projects. The historical perspective discussed by the PID Working Group is provided below.

Historical Perspective

The display of electronic information in the cockpit was first implemented in the 1960s with the introduction of navigation displays. As display and sensor technology improved, aircraft increasingly have been equipped with single purpose displays, multifunction displays, integrated electronic flight instrument systems, and now, integrated and networked information systems. The availability of more affordable and advanced displays and processors has now placed the aviation community at the threshold of a major technological advancement. As desktop, notebook and networked personal computers (PC) revolutionized industries throughout the world, this technology is now providing the aviation industry with great opportunities and challenges. The challenge is to bridge the gap between commercially available hardware and applications and demanding aircraft and operator requirements.

Some operators, attempting to capitalize on the opportunity to improve operations, reduce costs and improve safety through the introduction of an electronic document and information system, have focused on pieces of technology without a long-term vision and direction. Laptop PCs and industrial equipment have been evaluated and fielded with limited success. A few of these past and ongoing programs include:

- *United Airlines/NASA/Astronautics Pilot Access Terminal (PAT)* - As part of the Cockpit Weather Information Network (CWIN) program sponsored by NASA and United Airlines, a Pilot Access Terminal was developed by Astronautics and evaluated, using a DC-10, for a six month period. This was a landscape mode, PC based display with a 6" x 8" screen equipped with a touchscreen. PAT applications included: a GPS CDU, ACARS CDU, SATCOM CDU, TCAS and CWIN.

- *Boeing/Collins Side Display* - Boeing has provided a Side Display in a few B-777s as a Maintenance. This display was provided to three operators, however, it is only in service today with China Southern, due to its focus on maintenance functions and lack of functionality for flight crews.

- *Northwest/Aviontek Integrated Crew Information System (ICIS)* - An Integrated Crew Information System was developed in cooperation with Northwest Airlines to enable paperless cockpit operations. With the goal to minimize crew task loading and optimize human performance, the ICIS display was evaluated in the NASA 757 for the AWIN program, and a formal evaluation was performed in the Northwest Airlines Simulator. The ICIS facilitated the definition and development of many EFB concepts, but was never fielded due to its proprietary nature.

- *Federal Express/Spirent TITAN Project* - Federal Express has operationally fielded its TITAN project, a cooperative effort of Federal Express and Spirent. The system consists of the use of an Airborne File Server, a lap top computer based Pilot Access Terminal and ground based stations.

- *Southwest Airlines Performance Tool* - Southwest Airlines has utilized a Fujitsu 1200/2400 as a Performance Computer for Performance Calculations and MEL applications.

- *JetBlue EFB* - JetBlue has operated a system based on lap top computers with the philosophy being to only change the media and not the content or functions used in their fleet operations. They have implemented no changes to format, content or organization of the document system. Future plans do exist for implementation of gate connections and access to airborne information, if possible.

- *United Airlines Sky-Pad* - United Airlines has completed two phases of an EFB evaluation project as a follow on to the NASA AWIN project. The United Sky-Pad EFB provides access to Honeywell WINN, Jeppesen charts and Digital Flight Manuals in a United A-320. The system was based on the Fujitsu 3400 Pen Tablet, and relied on the United Airlines Pub Trax project to digitize operations documents.

- *Airbus 'Less Paper in the Cockpit' Tool* - Airbus has introduced a PC based 'Less Paper in the Cockpit' (LPC) tool focused on performance computations and consultation of technical information (FCOM/MEL). The LPC operates as a laptop computer that is stowed during take-off and landing. As the system is not certified, paper remains on board the aircraft and the hardware is not integrated with other aircraft systems.

- *Sobelair* - Sobelair has introduced an EFB in its 737 that utilizes the Boeing BLT Performance software, a PDF reader for company notes and Route Manuals with notes and pictures of the airports (using a JPG format developed in house). Their plan is to replace pilot-company communications (flight reports and notes) with an electronic format. The hardware used is the Fujistu Stylistic C500 LT with Obstacle database (for the Boeing BLT) and company documents on a compact flash card.

These efforts establish the potential for electronic information and confirm that further application system development is warranted. A review of these efforts has identified areas that need to be addressed in order to develop a successful EFB. The cockpit experiences a wide range of light levels and the display device must have a substantial brightness ratio to be usable at all times. A few of the displays met cockpit requirements, but were too expensive. On the software side, a robust, open source system is required to meet the higher levels of certification. These efforts also confirm that the display of information for all phases of flight requires the latest technologies to meet extensive cockpit and crew requirements.

Hardware Options

Relying on the field experience of the operators listed above, various hardware options were reviewed to determine what provides the most benefit to the operator. As illustrated in Figure 11.1, three basic cockpit display hardware configurations are available ranging from a certified, avionics grade display to a standard, commercial PC.

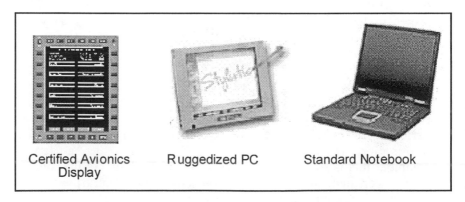

Certified Avionics　　　　Ruggedized PC　　　　Standard Notebook
　　　Display

Figure 11.1 Basic Cockpit Display Configurations

The pros and cons of each option were weighed and, although the commercial PC is low cost and can be treated as a throw away maintenance item, it is not a viable option for a full information management system. The cockpit demands for an extended brightness ratio, greater viewing angles, better reliability and maintenance supportability could not be met by low cost PCs. The higher end, ruggedized and sometimes tailored commercial PCs were demonstrated to the user group, and similar results were obtained. Although these ruggedized PCs are more reliable, they are still difficult to support because the hardware and operating system configurations change almost quarterly to meet commercial marketplace demands. This can result in having to support multiple configurations as well as a training challenge to maintain crew proficiency for different computer configurations. The need to enter data through an external keyboard or a stylist/touchscreen may be a further limitation.

One final drawback is that PCs are not hardware certified, so they must be stowed for take-off and landing, and the use of uncertified Microsoft® Windows® operating systems further limits the use of applications to Level D and no higher than Level C certification.

The preferred configuration of the working group participants was a certified avionics display that had similar optical qualities to other cockpit displays. The following benefits could be realized from such a display:

* The certified status would enable it to be used during all phases of flight.
* The ability to host a certified operating system would open the door for any future application programs, without limitation.
* The optical characteristics of the display can be tailored and optimized for cockpit requirements.
* This highly reliable type of display would be economically supportable throughout a large or small operator fleet.

Based on this industry guidance, the PID Working Group undertook to define what would be required of such a display. The display should have a brightness ratio of approximately 1500:1 or more, with preferred low-end luminance at 0.1 foot lamberts (fL) or below, and high end luminance of no less than 150 fL, with 200 fL desired. The display's viewing angles should be symmetrical for use at Captain and First Officer locations, with contrast ratios greater than 30:1 met at least to ±20° angles with viewing at greater angles also required. The display should have a minimum useable screen area of 8.4" diagonal with a 10.4" diagonal preferred. Additional display requirements include the ability to dim or increase screen brightness of the display. The display's colors, symbols and fonts should approximate those used on other cockpit displays. The user interface should be based on bezel controls to enable a task-oriented suite of applications with minimal key presses. The display should be certified to enable operation during take-off and landing and an upgrade path must be addressed.

Future Functional Requirements and Desired Applications

Having defined the hardware requirements for an EFB display, the next step is to determine what functional applications are required in the near term and desired for the future.

Significant issues to be addressed include the requirements for the quality, availability and certification level of an electronic information management display, which are closely based on which applications are desired. The types and information content of the potential applications have been extensively reviewed by working groups interested in the EFB, including the ATA's Digital Displays Working Group (DDWG), their Flight Operations Working Group (FOWG) and the NASA/FAA Operating Documents Group.

Based on the guidance and suggestions put forth by the working groups, an extensive list of potential application programs has been prepared, and industry teams are working on or are ready to field a number of applications for pilot and maintenance use. Proposed aviation information applications may include: Electronic Data Management System (EDMS), Aviation Weather INformation (AWIN), airport mapping, Terrain Awareness Warning System (TAWS), Traffic Information Service Broadcast (TIS-B), EFB, aircraft performance, Electronic CheckList (ECL), electronic aircraft log book, maintenance electronic documents, web browsers and numerous other aviation information applications.

Possible EFB applications are listed in Table 11.1. Of these applications, some most likely to be implemented near term include flight manuals, flight operations manuals, checklists, aircraft performance calculations and electronic charts. An example of a checklist application is pictured in Figure 11.2. In addition to operating aircraft information applications noted above, the display may function as a Controller Pilot Data Link Communication (CPDLC) and ACARS control display units, or any function commonly used on multifunction displays. The display can also provide access to a wireless digital information network in all phases of flight and on the ground.

Table 11.1 Possible EFB Applications

Flight Manuals	Normal Checklists	Crew Airport and Hotel Information
Flight Operations Manual	Aircraft Performance Calculations	Passenger Service Procedures
ACARS	Weight and Balance	Maintenance Manual
Controller to Pilot Data Link Communication	Aircraft Log Book	Flight Plans
Minimum Equipment List/ CDL Manual	Passenger List	Flight Planning
Emergency Checklists	Airline Operating Timetables	Jeppesen Airways Manual and Charts
Irregular Checklists	Security Procedures	

When reviewing the past working group research and perceived application requirements for an EFB, a recurring theme has been the desire to reduce or remove paper in the cockpit, reduce the related costs of procuring and maintaining updates and carrying the paper charts, checklists and manuals across large fleets. Despite this desire to reduce paper, several major operators commented that removing paper from the cockpit was not, by itself, sufficient to justify the investment in such an information management system. No accident or workload issues have been attributable directly to the presence of paper in the cockpit. Therefore, it was stated that some safety applications would also be required to make a business case for such a system.

As a result of industry input, a series of EFB related safety and security applications have been identified. Applications to improve situational awareness include satellite weather to assist turbulence avoidance, airport mapping to prevent runway incursions, enhanced vision and terrain warning. New security applications include video surveillance and flight crew alerting.

The Aviation Weather Information (AWIN) network, developed through the sponsorship of NASA, provides real time graphical weather information for the entire world via data link communication. Weather data provided from different sources is packaged as packets of data and relayed via a desired communication network node and then to the airplane. The data that is relayed represents changes to current and forecast conditions to assist the flight crew in improving its situation awareness. Digital data packets are designed for relay via HF, SATCOM, GateLink, VDL mode 2 and satellite telephone communication networks. AWIN provides a range of weather overlays, including satellite infrared, radar mosaic, convective, turbulence, winds, surface precipitation and volcanic ash.

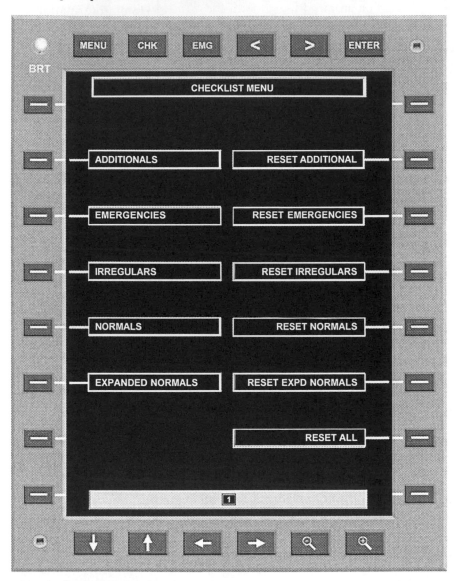

Figure 11.2 EFB Checklist Application

A second important safety development that has arisen from incidents attributable to runway incursions, such as recent crashes in

Taipei and Milan, is the display of airport ground maps for use with surface guidance and runway incursion systems. When combined with an FMS or, eventually, an ADS-B system, the position of the aircraft can be displayed on an airport map with the appropriate runways and taxiways highlighted. The assigned gate can also be highlighted.

Airport maps are planned to be a key element of an ADS-B implementation, and industry leaders Jeppesen and Honeywell are independently developing Runway Incursion Prevention/Airport Mapping applications relying respectively on the Jeppesen charts and Honeywell Enhanced Ground Proximity Warning System (EGPWS) Terrain Data Base to create airport maps. Enhanced or synthetic vision can also assist in preventing runway incursions. This information can then be hosted together with an information management system electronic display to improve situational awareness.

Another significant safety application available for integration on an electronic display system is a video surveillance system input. The terrorist actions of September 11th have stimulated industry to increase safety in the cockpits of aircraft. One method is to provide improved visual awareness to the flight crew of activities in the rear of the aircraft through the use of video cameras. Cameras placed in the aircraft cabin, cargo hold and wheel wells can also be integrated with an information management system to provide situation awareness input to the flight crew.

Human Factors

After a thorough review and evaluation of hardware and available applications, primary human factors issues were addressed to determine the optimum display of electronic information in the cockpit. These factors are critical in ensuring that crew workload is reduced when electronic information is introduced to the cockpit, and that safety is maintained by displaying accurate information that does not distract crews from other tasks.

The current draft of the EFB Advisory Circular (FAA, 2001) identifies the EFB's potential to increase safety by improving ease, speed and accuracy of information retrieval. It also identifies the need to provide the most current information along with the benefits of freeing up time for crews to perform other flight critical tasks. It addresses operation of an electronic library that contains reference material, or as an interactive device capable of providing a moving map with aircraft position, weather, traffic, terrain and obstacle data and aircraft system status, information currently provided in paper form.

A common human factors requirement across diverse user groups was the need to minimize the number of operator actions required to access applications and the desired information. Applications should be accessible via a menu structure that follows a 'top-down' menu architecture and allows the pilot to sequentially step through available pages. This menu architecture functionality is similar to operating a typical flight management computer or ACARS menu found on most commercial transport aircraft. When the MENU function key is selected, a list of available applications should be provided. Line selecting an available application selects the application or provides a subset page. An example includes line selecting the PERFORMANCE page that displays a subset performance page such as TAKEOFF, SPECIAL OPERATING GROSS WEIGHT, ENROUTE, ENGINE OUT, CRUISE, or LANDING performance page.

Various data entry methods have also been evaluated to minimize crew workload, leading to the following conclusions:

- The cockpit display should be configured with dedicated function and programmable keys similar to an FMS or ACARS CDU for ease of use.
- A touchscreen or external cursor control device can be provided, but should not be the primary means of data entry or mode selection.
- An external keyboard can be provided, but should not be the primary means of data entry or mode selection.
- Whenever possible, data entered through an existing cockpit system should be automatically input to the EFB display to avoid the duplicate entry of information. This can include data uploaded from the ACARS system and flight plans entered into the FMS.

Cockpit Location Considerations

The selection of a location for a display in the cockpit is also an important issue that should not be overlooked from a functional and human factors perspective. Basic certification guidelines, and common sense, mandate that new displays of this type should not interfere with existing controls, block the view of necessary data from an existing display or block egress from the aircraft. The display should also be installed such that it is readily accessible and viewable from all cockpit seat locations.

Readability of a chart or other display is critical, and must be maintained through a wide field of view of the EFB display when mounted in a fixed position, or a swivel mount must be provided for individual pilots to optimize the viewing angle. Various mounting configurations will have to be accommodated with new production B-777 aircraft having an installation location reserved for such a display, while 'classic' aircraft may require the moving of oxygen apparatus or replacing the PPI displays. For a system that does not require interfaces to a network or avionics systems, the goal is for overnight installation of the EFB display.

A review of operator cockpit configurations led to a conclusion that a single, common display location or mounting scheme in all aircraft types is not feasible. A flexible, modular mounting tray was the best approach to enable an operator to support a single display configuration with only the installation kit varying from aircraft type to aircraft type.

Certification

A critical element of displaying electronic information in the cockpit of a transport aircraft is the level of certification required. These hardware and software certification levels are closely tied to the type of application implemented in an electronic display system. To meet aircraft certification requirements for hardware, the display must comply with the following government and industry documents:

- FAR 25.1309 – Equipment, Systems and Installations.
- FAA AC 25-1309-1A - System Design Analysis.
- FAR 25.1301 - Function and Installation.
- FCC Rules Part 15 - Radio Frequency Devices.
- RTCA DO-160D - Environmental Conditions in Test Procedures for Airborne Equipment (RTCA).
- TSO-C113 – Airborne Multipurpose Electronic Displays.

The software must comply with the following industry and government documents:

- RTCA DO-178B - Software Considerations in Airborne Systems and Equipment Certification (RTCA, 1992).
- RTCA DO-255 - Avionics Computer Resource (RTCA, 2000a).
- RTCA DO-201A – Standards for Aeronautical Information (RTCA, 2000b).

Within the DO-178B software requirements (RTCA, 1992) there is the challenge of defining what level of certification is required for each application program. The major levels of certification are provided in Table 11.2. The type of application program and whether a backup paper copy is retained on the aircraft are the primary determining factors. Crew, passenger and scheduling documents, for example, would be Level E. Emergency Checklists operated on an Electronic Flight Bag without paper copies retained on the aircraft would be Level B. The majority of EFB applications would reside in the Level D and Level C. An interesting question related to certification requirements is, as electronic charts are not currently certified, does the use of their digital equivalent now require that they be certified? The answer at this time appears to be NO.

Table 11.2 Major Levels of Software Certification (RTCA, 1992)

Certification Level	EFB Considerations
Level A - Catastrophic Failure (Prevents safe flight).	No planned EFB applications.
Level B - Hazardous/severe Failure (Results in Severe/Hazardous condition).	Possibly emergency checklists.
Level C - Major Failure (Results in significant reduction in aircraft safety).	Several EFB applications.
Level D - Minor Failure (Results in minor reduction in aircraft safety).	Majority of EFB applications.
Level E - No Effect (Has no effect on safety).	Several EFB applications.

A greater challenge facing EFB providers is the level of certification of the operating system. Most EFBs are running with some variant of Microsoft Windows, which cannot be certified beyond Level D. Various manufacturers have or are attempting to certify application programs to Level C while operating in a pared down Level D operating system. Such an approach would limit the future growth of the electronic information management system to the lower levels of certification.

As a result, the PID Working Group and Astronautics elected to utilize a Linux based operating system that is being certified to higher levels to support all applications.

Conclusion

Developing, certifying and providing an EFB display as an integral part of an electronic information system can be a complex and dynamic task. A myriad of operation demands and opinions, FAA and manufacturer specification requirements, advisory circulars, working group inputs and human factors studies all play into the final equation of how to optimally display electronic information in the cockpit. The end result is achievable when the display specifications, control features, display location and the usability of the application programs all come together to provide a reliable, cost effective tool for use in cockpit applications.

Comparing the final PID product conceived and developed based on operator inputs, with the relevant NASA/FAA findings (see Chapter 1) listed in Table 11.3, finds that the desired end results appear to have been achieved.

Table 11.3 PID Compared with Top Rated Electronic Media Issues

Transition to Electronic Media Issue	Mean Rating	Achieved by PID
Take advantage of technology, don't use technology for technology's sake.	2.04	Yes
User interface/usability issue (e.g., hypertext).	2.07	Yes
Industry guidelines needed (e.g. SGML) media.	2.10	To be determined
End-user benefits.	2.14	Yes
Security.	2.17	Yes
Access/distribution.	2.18	Yes
Desynchronization of paper vs. electronic.	2.21	Yes
Training cost reduction.	2.29	Yes
Content guidelines remain same; presentation/format to optimize electronic media.	2.32	Yes
Certification and approval process unclear.	2.34	Process clarified
Revisions dates on electronic media.	2.38	Yes
Printing and distribution savings.	2.61	Yes
Weight/cost fuel reduction.	2.68	Yes

Following the road map outlined here can enable an operator to implement an electronic information system that can meet its near-term requirements, which are typically the most prominent, and also produce the more immediate and tangible results. At the same time, this approach can equally address the long-term concerns of the industry to establish a common strategic direction and lay the foundation for future growth and advancement that are sure to follow.

References

FAA (2001), *Operational Approval of Portable Flight Data Equipment – The Electronic Flight Bag* (AC 120-EFB), Washington, DC: Federal Aviation Administration.

RTCA (1992), *Software Considerations in Airborne Systems and Equipment Certification*, (RTCA DO-178B), Washington, DC: Radio Technical Commission for Aeronautics, Inc.

RTCA (1997), *Environmental Conditions in Test Procedures for Airborne Equipment*, (RTCA DO-160D), Washington, DC: Radio Technical Commission for Aeronautics, Inc.

RTCA (2000a), *Requirements Specification for Avionics Computer Resource*, (RTCA DO-255), Washington, DC: Radio Technical Commission for Aeronautics, Inc.

RTCA (2000b), *Standards for Aeronautical Information*, (RTCA/DO-201A), Washington, DC: Radio Technical Commission for Aeronautics, Inc.

VOLPE (2001), *Human Factors Considerations in the Design and Evaluation of Electronic Flight Bag: Version 2: Basic and Advanced Functions*, Cambridge, MA: Volpe National Transportation Systems Center.

Summary and Recommendations

Chapter 12

Future of Aviation Operational Information

Thomas L. Seamster and Barbara G. Kanki

Introduction

This book provides an overview of current practices and standardization efforts to make the creation, management and use of operational information more efficient. From these activities, we have outlined the vision, planning and implementation to help operators and industry make the critical transition from documents to data. Safety- and time-critical, human factors, security and standardization provide key underlying considerations as the industry migrates to electronic documents. In the absence of well established guidelines, these considerations and the recommendations they imply provide corporate and industry direction for improving the management and use of aviation information.

Economic realities also mandate cost/benefit considerations which should be thought of as necessary, but not sufficient, for developing a long-term vision and making near-term implementation decisions. Cost considerations are an essential part of the equation, but it is evident that for industry-wide efficiencies and savings, long-term benefits must be compared against near-term costs. We acknowledge the impact of near-term cost/benefits on planning and implementation, but emphasize the long-term benefits of industry-wide cooperation and standards in setting the goals and the vision.

Parts 1 and 2 of the book have dealt with the higher-level considerations of structuring and managing aviation operational information and are summarized in the next two sections on Integrating Information. In these sections we describe the current way operators structure, use and manage operational information, and how this process may change in the future. Beginning with a corporate perspective and the purview of individual operators and suppliers, we then review the changing approach to operational information from an industry perspective. Each point of view recognizes key issues pertaining to human factors, security and standardization but these issues have different implications and consequences for individual corporations versus the industry as a whole. With the higher level consideration in mind, we review the lessons learned from several current innovations in Part 3 and conclude recommendations for the direction and essential next steps.

Integrating Information: Corporate Considerations

The distinction between corporate and industry considerations is not always distinct and there is some overlap, but it is helpful to differentiate between what corporate entities can do individually and what they need to do as an industry group. For success, corporate and industry considerations have to align so that everyone is heading in the same direction. This direction is becoming clearer as operators make certain in-house investments and also participate in standards efforts that will shape the future of operational information. We first address operator concerns and then look at industry-wide directions in the next section.

Safety-Critical Considerations

Operators have been working for some time on restructuring their documents into an integrated system rather than a library of independent manuals. Early efforts focused on reducing the number or size of documents in order to accommodate space and weight constraints. The rationale for restructuring and consolidating document systems is a complex decision process resulting in structural tradeoffs with consideration of regulations and supplier recommendations, as well as human factors, usability and standardization guidelines. Tradeoffs typically resulted in compromises, except in the area of safety-critical information.

As shown in survey results from Workshop II (Kanki, Seamster, Lopez, Thomas and LeRoy, 2000), the top issues driving document system decisions were concerned with documents required in flight, that provide time-critical information. In essence, time-critical information is the primary safety-critical element. Also highlighted is the importance of standardizing abnormal procedure flows from one procedure or document to another with an effective indexing system. Another top issue was how to implement changes, including getting feedback from flight crews and others involved in the operation, testing organizational logic in the simulator under real time operation and developing communicating mechanisms for introducing new information to users. How to maintain standardization was also important, including standardizing procedures and flows and the use of style manuals, guides and master templates. These top issues involve human factors but the highest priorities go to that subset of documents that involve the retrieval of safety- and time-critical information. Recently, issues of document organization have combined with those of information architecture as operators are moving toward electronic document systems. Whether working with paper-based or electronic documents, operational needs for safety- and time-critical information should remain the top corporate considerations.

Human Factors Considerations

Existing approaches to developing document systems have utilized the support and expertise of user groups such as pilots or maintenance personnel at the procedural and document levels, but seldom at the overall document system level. Operators need to identify all information users and specify their shared information requirements. This is critical in the transition to electronic documents where current technology allows for the automated display of information tailored to specific users and screen formats.

Different user needs become highlighted when comparing human factor needs of those who manage documents publication versus those of the pilots who must access these documents in the cockpit. Those managing the documents work in an office environment with large displays and are focused on creating, editing and updating documents. Their needs revolve around a workflow that allows for the efficient management of these documents. They need to be able to view and track changes, make updates and have ways to check on information sources and accuracy. The cockpit provides a very different set of human factors requirements to address the extreme range of illumination, the time-critical and multi-tasking nature of the work and the team approach used by most operators. These are just two of the many different user groups that may ultimately access and use new electronic document systems. Corporations need to identify and then work with the different groups, making sure that the document systems meet the needs of all groups. Although this is one of the big advantages of a data based document system, where documents can be automatically repurposed for different uses, those different purposes need to be defined and understood in the planning phase. Depending on the corporation, pilots, engineers, information managers or publication experts may take the lead in designing the new system, and it is essential that this does not result in a system meeting the needs of the lead group at the expense of the other users. In this corporate move to electronic documents, the technology exists to address the human factors needs of all key user groups. What is needed is a corporate mandate that all groups will have their needs addressed equitably.

A second human factors consideration deals with the introduction of new technology, either hardware or software. As corporate entities develop electronic document systems, their associated technologies affect how documentation is managed in operations and how it can be used in the cockpit. The new technologies are very exciting, providing opportunities to automate many procedures and to improve information access and searching. Corporate entities need to concentrate on the appropriate use of these new technologies both in operations and the cockpit. Aviation has numerous examples of less than optimal technology implementations. For example, many early implementations of computer training were little more than electronic page turners where the technology was used to replace the function of training documents without taking advantage of a computer's interactive capabilities.

Similar misuses of electronic document technology are likely to occur, with some operators unknowingly under or over utilizing the technology. To follow through with the page turner analogy, some operators may simply take the documents, and without restructuring or reformatting, load them on computers to be used in the cockpit. There are obvious differences between paper and screens in their overall size, resolution and levels of contrast and these differences should be addressed when developing an electronic system. On the other end of the spectrum, in the area of technology overuse, computers offer a wide range of functionality, some of which may not be appropriate for use in the cockpit. Electronic documents, especially those within a well designed document system, offer tremendous improvements that will only come about through careful attention to human factors considerations.

Historically, the regulatory approval and certification process have been equipment- and technology-centric, and even with some of the recent guidance such as FAA (1996), human factors is implemented more as an overlay rather than being fully integrated into the process. Even the human factors guidance under development for the Electronic Flight Bag (EFB) categorizes requirements and recommendations by equipment, installation and training/procedures (Volpe, 2001) with user-computer interface issues located several layers below. It is up to the operators and suppliers to take the lead as regulatory guidance catches up with technology and best practices.

With neither well established guidance nor fully integrated human factors procedures, operators and suppliers must ensure that new electronic document technologies are properly evaluated and tested. When such testing is undertaken, operators should require that it be conducted over the full range of normal and non-normal conditions, including worst-case scenarios. Boorman (2000), in his research on electronic checklists (ECL), points to the need for a consistent user interface between the frequently used parts (e.g., normal items) and infrequently used parts (e.g., non-normal items). Implied in this research is the caution against designing on the basis of normal, frequent events alone. For example, if a particular EFB function requires pilot decisions during several non-normal conditions, but can be automated for most other cases, non-automation may be the optimal choice. If you automate the function for 95 per cent of the time, but require pilot decision for even a small number of critical situations, pilots may not have maintained the required skills to make a quick and accurate decision. If it is possible to develop a greater consistency between normals and non-normals through less automation, that option should be strongly considered.

Security Considerations

Electronic document system technologies offer flexibility in the access to and distribution of operational information, thus increasing the need for greater security. As a top aviation priority, operators should not only concentrate on security of the physical aircraft and airport; rather, this is the time to look at the security of all information assets to include operational information within a comprehensive security plan. In 1992, the Office of Technology Assessment (OTA, 1992) warned that as security, legislation and enforcement measures increase, terrorists are likely to attempt more 'daring' acts such as the identification of 'vulnerable mass targets'. The concept of vulnerable mass targets has become a reality, and with the increase of aviation security, the industry must be on the alert to guard against physical and electronic security risks.

During the development of a new electronic document system, corporations need to review all security aspects, including who has access to the information, how that access is controlled and level of security of the system to include servers, networks and all entry points. System developers need to guard document system content from unauthorized access without making it difficult for authorized users to gain access. In addition, developers must implement procedures to ensure content integrity for operational information that may undergo numerous updates and may be distributed in multiple forms.

Corporations need to address all these levels of security across the entire electronic document system. Further, a document system must have safeguards to prohibit tampering and inadvertent changes to operational information. Security cannot be ensured just through technology. Security solutions involve the careful selection and implementation of technology, balanced with a corporate security plan and procedures tailored to its mission and personnel. The human element must be at the core of all technology and procedures. Technical solutions such as firewalls, content security software and security probes help, but analyses show that many security incidents result from problems with security policy and its enforcement (NASA/FAA, 2000). The human element must be addressed in the design of security policies and procedures. If security adds needlessly to workload, does not yield results, or causes delays, personnel will find ways to circumvent the system, often introducing even greater risks. A familiar example in operational environments is the slow responding code or card entry system that encourages tailgating. If the security system does not have adequate response times, breaks down frequently, or is not integrated with other operator tasks, it will likely be circumvented. At its core, security is only as good as the corporate security policy implementation and adherence by employees. Security needs to be actively integrated in corporate philosophy, policy and procedures that everyone understands, supports and follows.

Standardization Considerations

Corporate standardization of operational information is essential, and although that standardization does not have to match industry standards, it should be compatible. Corporate divisions, like flight operations, have different departments such as flight standards and training. These departments use some of the same information, shared information, in their documents and training materials. Corporations need to identify existing shared information as well as additional information that could be shared. At the higher corporate level, there is shared information about the aircraft, crews and logistics. Divisions like flight operations, in-flight services, logistics and maintenance need to identify areas of duplication and sources of shared information. Beyond electronic documents systems, an overarching data system that identifies areas of standardization will add to corporate efficiency.

There is a strong need for both guidelines and standards in this new area of electronic documents. With paper documents, operators had access to substantial guidance from the research community and other operators' best practices. In this new domain of electronic documents there is little guidance, and most that exists is still under development. Standardization is further complicated when an operator is working with both paper and electronic documentation. Guidance can be ambiguous because they may suggest that operators maintain a certain level of compatibility with paper documents while advising that they address the unique human factors issues of electronic documentation at the same time. The need for more guidance is essential, but with limited studies and few opportunities for data collection, it will take time to develop solid guidance based on actual flight conditions and experience.

Just a few years ago, standardization had a relatively parochial meaning for most operators. When we talked about standards and standardization, our vision focused mostly at the within and across fleet level (NASA/FAA, 1997). While recent work on industry-wide standards (ATA, 2001) has broadened the concept beyond the fleet environment (as discussed in the next section), corporate standards have not fully expanded across department and division boundaries within organizations. Much work remains in identifying shared information user groups and in developing an integrated information data base. Even the smaller operators whose single or dual fleet standardization process is relatively simple may gain significant benefit from repurposing information data for multiple users.

Integrating Information: Industry-Wide Considerations

The aviation industry must work together on the superordinate considerations from a broader perspective, taking into account not only the corporate, but also the regulatory environment. Issues of safety-critical, human factors, security and standardization all play a part, with standardization becoming the main emphasis of the industry effort. A new level of cooperation is needed between operators, suppliers and regulators to realize the full benefits of data interchange standards. Each of these three industry entities constantly exchange data with the other, concentrating on their own requirements or, in the case of regulators, mandating certain standards, without sufficient consideration of industry-wide direction and needs.

The aviation industry should concentrate on a cooperative, phased approach to standards development. This approach would maintain a long-term direction and vision but would carefully define, specify and implement the foundation of the standard in the near term. This phased approach should be based on an iterative development cycle gradually introducing new data entities as additional groups are brought into the data interchange endeavor. The Air Transport Association (ATA) Technical Information and Communication Committee (TICC) has started with aviation maintenance and logistics, and has brought in flight operations with the work of the Flight Operations Working Group (FOWG). They have concentrated on supplier to operator data interchange. Their existing standard needs to be implemented and then expanded to other aviation industry information interchanges that include regulators.

One example of a cooperative, phased approach for standards development is starting to evolve through the work of the ATA FOWG. Under that effort, operators and suppliers have developed a data model that accounts for the main data interchange entities or information objects. Operational information is complex and the FOWG data model specifies the relationship between the primary objects (see Figure 12.1 for a simplified view of the model). This model includes those objects central to pilot performance; action, annunciation, environment, phase of flight and procedure. The other information objects, or entities, are more closely related to aircraft performance, including aircraft state, aircraft system, limitation, malfunction, MMEL dispatch, planning information and maintenance task.

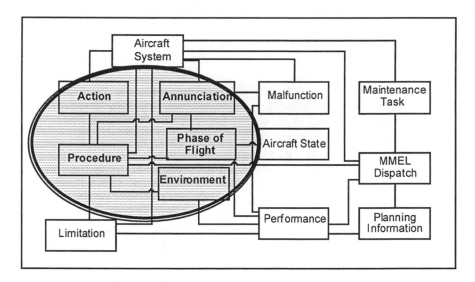

Figure 12.1 Simplified FOWG Data Model Emphasizing Pilot Performance Entities

A closer look at this model reveals that the entity, aircraft system, is a primary link between pilot performance entities and aircraft performance entities. This flight operations data model is at a critical stage that requires harmonization with maintenance. Some of the maintenance entities are undergoing review by other TICC working groups, and now is the time for flight operations and maintenance to join forces and ensure that their models are compatible. The ATA and its working groups need to demonstrate the advantages of implementing these models into a data interchange standard.

The FOWG mandate is to implement an open source data interchange standard. They have developed some supporting business cases and are developing a proof of concept based on the most established and critical entities of the model. Phase of flight is one of those entities recently established as a standard and central to pilot performance. Aircraft system is a second entity because it is also a standard and a key link between pilot and aircraft performance. Several other entities, including annunciation, procedure and MMEL dispatch should be considered for this initial proof of concept. Operators and suppliers should jointly work out the proof of concept for some of these entities and demonstrate to industry how such an approach will expedite information interchange and facilitate overall information management.

The pragmatic approach being proposed combines two concepts, a phasing in of standards with frequency of updates. First, the phased approach to standards development advocates an iterative implementation based on several phases. Rather than spending five years developing one standard that accounts for all information objects, develop them in phases leading up to a comprehensive standard. Start with the best defined information objects such as 'phase of flight' and 'aircraft system', adding other objects in later phases. Second, frequency of updates suggests that those objects most frequently updated are the best candidates for the early phases of the data interchange standard. Thus, phasing and frequency of update combine to help industry identify and prioritize the information objects leading to integrated sets of standards built and tested in phases gradually over time.

Other research warns of the difficulties in achieving a simple decomposition of operational information into categories such as procedures and air transport systems (Blomberg et al., 2000). FCOM data is complex, and multiple levels of detail can be identified even within a single sentence. The phased approach removes the necessity of decomposing all operational documents prior to establishing a data interchange standard. The complete decomposition of the FCOM is many years, or several phases, off into the future. The industry needs results today or at the very least in the near future. With the phased approach that standardizes key pieces of operational information, the task becomes more manageable, especially if practical criteria like degree of existing agreement and update frequency are used to select the entities.

When reviewing the ATA information standards effort (ATA, 2001), the regulatory side seems all but absent. The FAA, although not a direct participant in ATA efforts, has been working on its own agency-wide information technology standards. That work shares many of the same goals with the ATA standardization effort to 'make reliable information available quickly' (FAA, 1999). The regulatory side is focusing on the National Airspace System (NAS) while the ATA is focusing on the operator to supplier business environment. These two domains have a large quantity of shared information. The phase of flight is just one of many information objects that needs to be implemented as a standard (Travers, 2000). Regulatory agency-wide standardization efforts need to be aligned with industry-wide efforts for a more efficient data interchange. Industry standardization is hard work, but it is essential that all players, including suppliers, operators and regulators, cooperate to ensure a true industry standard for information interchange. This gives operators a critical role that involves both championing a vision of an industry information interchange standard and remaining an active participant in its development. Once industry has succeeded in outlining such a standard, individual operators will benefit from the direction that an information interchange provides. They will be able to better select document conversion formats and workflow processes to address internal corporate needs while conforming with the industry standard.

   Both industry and individual corporations need to act cooperatively in order to progress. Standards for tomorrow are being developed today, and those not participating in the process may find it difficult to implement new standards.

Lessons Learned and Recommendations

As we indicated in the introduction, the transition from documents to data is just taking shape and the long-term direction is easier to define than immediate next steps. Returning to our main considerations of safety-critical, human factors, security and standardization, we would like to present a view of the future of aviation operational information that is optimistic, but constrained by some of the present realities of the aviation industry.

Safety-Critical Recommendations

As discussed earlier, industry has given special status to operating documents that contain safety- and time-critical information and this priority is extended throughout all phases of the design, development, production and maintenance of paper-based or electronic documents. In this area, at least at the top levels, there appears to be consensus across not only operators and their suppliers, but the regulators as well, resulting in the regulations and recommendations that operators consider first. In spite of this unifying agreement, the way in which safety- and time-critical information is made accessible, unambiguous, up-to-date and human centered is left up to the regional FAA, document developers, flight standards and ultimately, the users. In light of little validated guidance and limited experience, current innovations in developing an EFB or parts of an integrated information network have proceeded cautiously, often increasing redundancies and backups while testing new technologies. While the safety-critical priorities are the same across paper-based and electronic systems, the way in which these priorities are addressed may be quite different and as yet untested:

* Safety-critical information holds top priority in document design, development, production and maintenance decisions regardless of media.
* Safety- and time-critical information should involve high levels of user participation at design and testing phases.
* Safety- and time-critical information requires protected and secure status with respect to its place in the information database and management system.
* The way in which safety-critical information is managed is closely tied to effective incorporation of human factors requirements and standardization policies.

The effective review of human factors and standardization issues related to the use of safety-critical information such as time-critical access and retrieval of information, navigation from one procedure to another, and transitions across normal and non-normal procedures, is largely dependent on how well users are integrated into the design and testing of procedures and new technologies. In theory, this is not different from the way all information structure and management should be addressed; it simply holds higher priorities in the decision process and may take precedent when tradeoffs must be made.

Human Factors Recommendations

If in the rush to implement, human factors are not fully evaluated, the negative impact on pilot performance can be significant and long lasting. Yes, we can train around these problems, as we have attempted to do in the past, but the long-term costs outweigh any short-term savings garnered at the expense of human factors. User interface, especially in the computer industry, is technology-centered (Grudin, 1990) and from an engineering perspective may be thought of as just one of the many system interfaces. In fact, the system components of the user interface, to include the input and display devices, are only a small part of the full user interface that includes documentation, procedures, training and support that shape how individuals actually use a system (Grudin, 1990). In the area of human factors, there are several main considerations to be addressed:

- Need for infrastructure that moves toward shared information.
- Multi-user group involvement throughout the process.
- Appropriate use of new technology.
- Development of integrated guidelines that satisfy HF evaluation and testing.

We may discuss infrastructure at both corporate and industry levels, but in both cases, there is a need to move toward 'sharing' information. At the corporate level, there are not only organizational boundaries that must be crossed, but legacy documents that are cumbersome to change. Support for identifying existing shared information and for discovering new areas of shared information take the form of corporate philosophy, policies and open communication and collaboration among departments and divisions within an organization. In essence, what were once independent groups or islands must evolve into one system that encourages a single source of information and the standardization of style guides, updates, software, tracking systems and approval processes among suppliers, operators and regulators.

We have seen several cases of multi-user involvement that begin with the creation of a 'core team' who hold the vision and planning responsibility for the long-term transition to an integrated information system. They in turn begin to identify the specific user groups who must be involved in the development process, from defining user requirements in operational domains (e.g., pilots, maintainers, planners and schedulers) to developing publishing standards (e.g., writers and editors) and making the information technology (IT) decisions for converting static to structured documents (e.g., markup language experts, IT programmers, business process stakeholders). In addition to actual information content and usability requirements (where, when and how information must be used) within user groups, there must be focused attention to group interfaces. For instance, when flight operations and maintenance collaborate via an electronic logbook.

During the development phase, a crucial role of multi-user groups is to perform usability testing at the procedural level to ensure correctness and fidelity of a procedure, particularly for safety- and time-critical information. Also of particular importance is the usability testing performed for new technologies. While it is sometimes difficult to resist the powerful and automated capabilities of new hardware, software and management tools, it is critical to use new technologies appropriately, conducting user testing at the interface level as well as under realistic operational conditions so that workload and transfer tasks can be assessed. Further, operational conditions that represent infrequent, non-normal, as well as frequent normal, must be considered so that inappropriate over generalizing is avoided.

Continuing user involvement throughout the development process includes the considerations of training and the skill levels your user groups must attain. To avoid lengthy and difficult skill acquisition, several operators have made design decisions that keep training demands to a minimum. In addition to technical skills and training, users must 'accept' the technology and, in addition to designing in communication mechanisms for updating and future plans, user acceptance can be facilitated by integrating the use of technologies across operational areas. For instance, one operator discussed the advantage of enhancing transfer of skills by using the same computers in training as the ones used in flight operations.

A final consideration is the development of integrated guidelines that satisfy human factors evaluation and testing so that user requirements and standards can be captured and updated throughout the development process and beyond. For instance, the process for incorporating user information must be ongoing because issues may emanate from any part of the process from hardware reliability to information management issues due to new regulations or acquisition of new aircraft. There must be a flexible process so that changes or modifications can be made in a phased or incremental process. Until this new approach to information structure and management has acquired a history of use and standards for development, a phased approach will be an effective means to test and verify that user needs are met.

Security Recommendations

With security becoming a top priority within aviation, recommendations are drawn from related industries in the context of new aviation risks and threats. Essential information is vulnerable to two forms of attack, physical and virtual. The physical information environment includes access to computers, storage and networks that house operational information. The virtual environment includes all other forms of access, including on-line and wireless. Security is a complex topic with many levels being addressed by a wide range of sophisticated technology, but at its foundation there are a few key considerations:

- Security programs should be corporate-wide, involving all systems and all individuals.
- Security is a dynamic process based on continuous probing and improving.
- Security addresses both physical and virtual assets.
- Security needs to address all levels of the document system.

Security programs must be planned and implemented corporate-wide, addressing both the technology and human element. Security must be tailored to the mission, goals and personnel of the corporation. Therefore, planning for the security program should include representative user groups and should emphasize the human element. The program should be integrated with the organization at all levels, starting with corporate philosophy and following through to the procedures. The program must be supported at all levels, especially the top levels of the organization; and the program must include procedures for recovery when security has been compromised or information destroyed or maliciously altered.

Security is dynamic, requiring constant testing and improvements. This is particularly important in the aviation environment where procedures and adherence to procedures has played such an important role. Instituting fixed security procedures does not work. Once fixed, security procedures can be exploited and quickly fail to address evolving or new security threats. Therefore, a security program must include extensive testing and probing of both the physical and virtual environment. Keeping up with new threats and constant testing are essential to a secure system.

Security needs to address access to physical devices as well as virtual information access. Operators have focused on physical security on aircraft and at airports, important factors for a safe and secure environment. Equal attention must be given to virtual security, including on-line and wireless access, as well as the ongoing validity of the data. Automatic checks of data validity may be one of the strongest near-term arguments for transitioning from documents to data. Validity checks for structured data, such as XML markup, are substantially more robust and flexible that the limited file-compare routines available to word processing document files.

Security of the virtual environment must be established at all levels of the information system. This includes the network, the operating systems for the electronic document system with its displays and all applications running under those systems. This is an ongoing process where organizations need to be aware of new outside threats, like those to operating systems, as well as new risks brought about by modifications or updates to applications and modifications to networks.

The most important lesson learned from recent terrorist attacks is to expect the unexpected and to imagine the unimaginable.

Standardization Recommendations

Standardization has both a corporate element and an industry-wide element, and both are essential to more efficient creation, management and interchange of operational information. The main recommendations are:

* Standardization of operational information should aim for the efficient interchange of data between operators, suppliers and regulators.
* Standardization should be implemented through a series of effective phases, establishing proof of concept within each phase.
* Standardization should be based on long-term savings and proof of concept.
* Standardization needs to be synchronized between the corporate and industry level.

The aviation industry should aim for nothing less than efficient data interchange between operators, suppliers and regulators. The ATA has been working on an operator with supplier interchange standard, and that needs to be expanded to encompass the regulatory community. A higher level of cooperation is needed between all parties to identify shared data and then develop the standards that will allow for efficient data interchange.

Standardization, at the corporate and, especially at the industry level, should be based on a phased approach. Such an approach allows the agreed upon elements to be standardized and serve as a stable base for the next phase. In this new arena of electronic documents, with rapidly changing technologies, it is better to make phasing decisions on data objects or data model entities rather than on specific technologies. Such interchange standards can provide longer-term and more stable guidance than specific coding solutions such as HTML or SGML.

A key to a successful phased standardization is developing a proof of concept at each phase. Such a proof of concept not only tests the standards, but also should demonstrate the economic advantages for adopting the standard. Operators and suppliers should demand such proof of concept and should be willing contributors to and participants in such efforts.

There must be a strong link between corporate and industry standardization, especially in the area of information exchange. Operators and suppliers need to be participants at the industry level, forging standards based on industry-wide needs rather than limited to self interests. Corporations benefit from this participation by becoming better informed about the directions that standards are taking and learning from other operators' best practices. It is a combination of these directions and best practices that should be used by operators and suppliers to make the difficult decisions of the what and when to implement specific elements of their electronic document system. Operators should plan strategically for their electronic document system, but they should implement operationally.

References

ATA (2001), *iSpec 2200: Information Standards for Aviation Maintenance* (CD-ROM), Washington, DC: Air Transport Association.

Boorman, D.J. (2000), 'Reducing flight crew errors and minimizing new error modes with electronic checklists', *Proceedings of the International Conference on Human-Computer Interaction in Aeronautics,* pp. 57-63, Toulouse: Cepadues-Editions.

FAA (1996), *Human Factors Design Guide for Acquisition of Commercial-Off-The-Shelf Subsystems, Non Developmental Items, and Developmental Systems,* DOT/FAA/CT-96/1, Atlantic City, NJ: FAA Technical Center.

FAA (1999), *Information Technology Strategy, FY2000-FY2002,* Version 1.0, Washington, DC: Federal Aviation Administration.

Grudin, J. (1990), 'Interface', *Proceedings of the Conference on Computer-Supported Cooperative Work,* pp. 269-278.

Kanki, B.G., Seamster, T.L., Lopez, M., Thomas, R. J., and LeRoy, W.W. (2000), 'Design and use of operating documents', In Appendix A, *Developing Operating Documents Manual, Moffett Field,* CA: NASA Ames Research Center.

NASA/FAA (1997), *Proceedings of the NASA/FAA Operating Documents Workshop II:* September 10-11, 1997, Dallas/Forth Worth Airport: American Airlines Flight Academy.

NASA/FAA (2000), *Developing Operating Documents Manuals,* Moffett Field, CA: NASA Ames Research Center.

OTA (1992), *Technology Against Terrorism: Structuring Security,* Washington, DC: Congress of the United States, Office of Technology Assessment.

Travers, R.W. (2000), 'Pilot Centered Phase of Flight Standardization', *Proceedings of International Conference on Human Computer Interaction in Aeronautics,* pp. 51-56, Toulouse France, Cepadues Editions.

VOLPE (2001), *Human Factors Considerations in the Design and Evaluation of Electronic Flight Bags (EFBs): Version 2: Basic and Advanced Functions,* Cambridge, MA: Volpe National Transportation Systems Center.

Index